Ingo Kowarik

Stefan Körner

Wild Urban Woodlands

New Perspectives for Urban Forestry

Ingo Kowarik
Stefan Körner
(Editors)

Wild Urban Woodlands

New Perspectives for Urban Forestry

With 107 Figures

Professor Dr. Ingo Kowarik
Technical University Berlin, Institute of Ecology
Rothenburgstraße 12, 12165 Berlin
Germany
e-mail: kowarik@tu-berlin.de

Dr. Stefan Körner
Technical University Berlin, Institute of Ecology
Rothenburgstraße 12, 12165 Berlin
Germany
e-mail: stefan.koerner@tu-berlin.de

Library of Congress Control Number: 2004115127

ISBN 3-540-23912-X **Springer Berlin Heidelberg New York**

Springer is a part of Springer Science+Business Media
springeronline.com

Printed in Germany

Cover design: Erich Kirchner
Production: Luisa Tonarelli
Layout: Camera-ready by Dr. Uwe Starfinger, Berlin
Printing: Mercedes Druck, Berlin
Binding: Stein + Lehmann, Berlin

Printed on acid-free paper 30/2132/LT – 5 4 3 2 1 0

Preface

The outstanding social and ecological roles of urban forests in the growth of cities has become widely known. In many parts of the world, despite or even because of continuing suburbanization, initiatives are being put forth to preserve urban forests, to develop them further and to make them accessible to the public.

This volume focuses on a particular component of the urban forest matrix – urban wild woodlands. We understand these to be stands of woody plants, within the impact area of cities, whose form is characterized by trees and in which a large leeway for natural processes makes possible a convergence toward wilderness. The wilderness character of these urban woodlands can vary greatly. We differentiate between two kinds of wilderness. The "old wilderness" is the traditional one; it may return slowly to woodland areas when forestry use has been abandoned. The enhancement of wilderness is a task already demanded of urban and peri-urban forestry in many places.

This book would like to direct the attention of the reader to a second kind of wilderness, which we call "new wilderness." This arises on heavily altered urban-industrial areas where abandonment of use makes such change possible. The wild nature of urban abandoned areas was discovered in the 1970s through urban-ecological research. Since then, in a very short time, profound structural changes in industrial countries have led to hundreds or thousands of hectares in urbanized areas becoming available for natural colonization processes.

This leads to the paradoxical situation that cities continue to grow nearly unchecked, while at the same time large expanses of land in the centers of shrinking cities are no longer needed, and in principle, are available for the development of spontaneous nature. Areas particularly affected in this way are those which have been shaped by industry and for which no future prospects for mining, steel working, etc. exist.

The articles in this book make clear that wild urban woodlands offer great opportunities for providing ecological and social functions to urban residents. However, new wilderness is not easily accepted by a broad public. It emerges as an immediate neighbor in heavily developed areas, but is frequently a *terra incognita* to which humans have developed no entry.

For nature conservation, forestry, open-space planning and landscape architecture, new wilderness means new demands as well. Does the nature of abandoned areas correspond to the nature conservation model that is usually oriented toward traditional pristine ecosystems or toward the remnants of the pre-industrial cultural landscape? What role can forestry play on sites where no forester has ever been? Can open-space planning make use of sites whose character is completely different from conventional green spaces and is instead shaped by the remnants of earlier urban-industrial uses and uncontrolled ecosystem dynamics? Is design by landscape architects superfluous or is it, on the contrary, necessary, in order to facilitate access to the sites?

For the old wilderness as well as the new, the same is true: A sustainable development of wild urban woodlands, a wide acceptance of the opportunities for further development of urban spaces that are connected with this, will only succeed through an alliance between different disciplinary perspectives. The contributions to this book address this topic from social- and natural-scientific perspectives and lead to integrated conceptual approaches.

The articles in the first section characterize wild urban woodlands as a new component of urban forests. In the second section, attitudes toward wild woodlands are the focus. Following that, ecological studies provide an understanding of natural processes in urban woodlands. In the fourth and final section, the integration of different perspectives within the framework of conceptual approaches follows, illustrated with concrete projects.

The contributions are the result of the international conference "Wild Forests in the City – Post-industrial Urban Landscapes of Tomorrow." The conference took place from 16–18 October 2003 in Dortmund, Germany and was organized by the Institute of Ecology at the Technical University Berlin in cooperation with the *Projekt Industriewald Ruhrgebiet* (Industrial Forest Project of the Ruhr).

To the many who were involved in making the conference and this book a reality, we would like to offer our sincere thanks. Thank you, first of all, to Thomas Neiss, chairman of the advisory board to the *Industriewald Ruhrgebiet*, for providing the initiative for the conference. The conference and this publication were supported by funds from the federal state of North Rhine-Westphalia through the State Development Corporation of North Rhine-Westphalia. With the organization of the conference, Renate Späth of the Ministry of the Environment and Conservation, Agriculture and Consumer Protection of the State of North Rhine-Westphalia as well as Michael Börth and Oliver Balke of the *Projekt Industriewald Ruhrgebiet* provided enormous support. From the large group of our Berlin supporters, we would especially like to thank Lorenz Poggendorf and Gisela Falk who ensured that the Dortmund conference was perfectly organized. Finally we would like to thank the authors for their contributions and the colleagues who assisted in the review process. And last but not least, for the realization of this book in its present form, we would like to thank Uwe Starfinger, who undertook the layout and technical editing, and Kelaine Vargas, who translated many of the book's chapters and provided the language editing.

Overall we hope that this book will contribute to furthering opportunities for the development of wild urban forests as a special component of the urban forest matrix.

Berlin, October 2004

Ingo Kowarik and Stefan Körner

Contents

Ecological studies

Conceptual approaches and projects

List of Contributors

Bauer, Nicole, Dr
Swiss Federal Research Institute WSL, Section Landscape and Society
Zürcherstrasse 111, 8903 Birmensdorf
Switzerland
E-mail: bauer@wsl.ch

Bell, Simon
OPENspace Research Centre, Edinburgh College of Art
79 Grassmarket, Edinburgh, EH1 2HJ
United Kingdom
E-mail: s.bell@eca.ac.uk

Burghardt, Wolfgang, Prof Dr
Faculty of Biology and Geography, University of Duisburg-Essen
Universitätsstr. 5, 45117 Essen
Germany

Davies, Clive
North East Community Forests
Whickham Thorns, Market Lane, Dunston, NE11 9NX, Tyne and Wear
United Kingdom
E-mail: clive.davies@necf.org.uk

Dettmar, Jörg, Professor Dr-Ing
Design and Landscape Architecture, Department of Architecture
TU Darmstadt
El-Lissitzky-Str. 1, 64287 Darmstadt
Germany
E-mail: dettmar@freiraum.tu-darmstadt.de

Drexler, Justina, Dipl-Ing Landschaftsarchitektur
Schulstr. 5, 82288 Kottgeisering
Germany
E-mail: justina_drexler@web.de

Dunnett, Nigel, Dr
Department of Landscape, Sheffield University
Arts Tower, Western Bank, Sheffield S10 2TN
United Kingdom
E-mail: n.dunnett@shef.ac.uk

Gausmann, Peter
Department of Biology, Ruhr-Universität Bochum
Universitätsstr. 150, 44780 Bochum
Germany

Grosse-Bächle, Lucia, Dr
Am Deichfeld 10
30890 Barsinghausen
Germany
E-mail: grossebaechle@web.de

Haag, Rita
North Rhine-Westphalia State Environment Agency (LUA NRW)
Wallneyer Str. 6, 45133 Essen
Germany

Haeupler, Henning, Prof Dr
Department of Biology, Ruhr-Universität Bochum
Universitätsstr. 150, 44780 Bochum
Germany

Hamann, Michael
Büro Hamann & Schulte
Koloniestr. 16, 45897 Gelsenkirchen
Germany

Henne, Sigurd, Dipl-Ing
mühlinghaus + henne, BfL Mühlinghaus Planungsgesellschaft mbH
Bahnhofstraße 13, 64625 Bensheim
Germany
E-mail: bfl.bh@t-online.de

Hitchmough, James, Dr
Department of Landscape, Sheffield University
Arts Tower, Western Bank, Sheffield S10 2TN
United Kingdom
E-mail: j.d.hitchmough@shef.ac.uk

Ichinose, Tomohiro, Dr
Institute of Natural and Environmental Sciences
University of Hyogo
954-2 Nojimatokiwa, Hokudan-cho, Tsuna-gun, Hyogo 656-1726
Japan
E-mail: ichinose@gakushikai.jp

Jorgensen, Anna, Dr
Department of Landscape, Sheffield University
Arts Tower, Western Bank, Sheffield S10 2TN
United Kingdom
E-mail: a.jorgensen@sheffield.ac.uk

Keil, Andreas, Dr phil
Institute of Geography and Didactics of Geography
University of Dortmund
Emil-Figge-Str. 50, 44227 Dortmund
Germany
E-mail: andreas.keil@uni-dortmund.de

Konijnendijk, Cecil C, DSc (Agr & For) ir
(formerly Danish Centre for Forest, Landscape and Planning)
woodSCAPE Consult
Rytterager 74
DK-2791 Dragoer
Denmark
E-mail: woodscape@mail.dk

Körner, Stefan, Dr
Institute of Ecology, TU Berlin
Rothenburgstr. 12, 12165 Berlin
Germany
E-mail: Stefan.koerner@tu-berlin.de

Kowarik, Ingo, Prof Dr
Institute of Ecology, TU Berlin
Rothenburgstr. 12, 12165 Berlin
Germany
E-mail: kowarik@tu-berlin.de

Langer, Andreas Dr
Planland
Pohlstraße 58, 10785 Berlin
Germany
E-mail: a.langer@planland.de

Leder, Bertram
The Land Institute for Ecology, Soil Policy and Forests
North Rhine-Westphalia (LÖBF)
Herbreme 2, 59821 Arnsberg
Germany

Lorenz, Antje, Dipl-Ing
Vegetation Science and Landscape Ecology
University of Applied Sciences Anhalt
Strenzfelder Allee 28, 06406 Bernburg
Germany
E-mail: alorenz@loel.hs-anhalt.de

Ono, Ryohei, Prof Dr
Graduate School of Agricultural and Life Sciences, University of Tokyo
Yayoi 1-1-1, Bunkyo-ku, 113-8657 Tokyo
Japan
E-mail: ono@fr.a.u-tokyo.ac.jp

Rink, Dieter, Prof Dr
Department of Economy, Sociology and Law
UFZ Centre for Environmental Research Leipzig-Halle
Permoserstr. 15, 04318 Leipzig
Germany
E-mail: rink@alok.ufz.de

Schulte, Annette
Büro Hamann & Schulte
Koloniestr. 16, 45897 Gelsenkirchen
Germany

Stempelmann, Ingrid
North Rhine-Westphalia State Environment Agency (LUA NRW)
Wallneyer Str. 6, 45133 Essen
Germany

Tischew, Sabine, Prof Dr
Vegetation Science and Landscape Ecology
University of Applied Sciences Anhalt
Strenzfelder Allee 28, 06406 Bernburg
Germany
E-mail: tischew@loel.hs-anhalt.de

Weiss, Joachim, Dr
The Land Institute for Ecology, Soil Policy and Forests
North Rhine-Westphalia (LÖBF)
Castroper Str. 30, 45665 Recklinghausen
Germany
E-mail: joachim.weiss@loebf.nrw.de

Wild Urban Woodlands: Towards a Conceptual Framework

Ingo Kowarik

Institute of Ecology, Technical University Berlin

New woodlands as a response to social and economic changes

Since the Neolithic Revolution, a decline in pristine forests has occurred in Europe. Around 750 AD Germany was still approximately 90% covered by forest. The growth in agriculture and the wave of cities being founded led to the intense clearing of forests at a rate never before experienced. Only a few centuries later, in the late 13th century, the greatest extent of deforestation in Germany was reached, with forest cover of only 17%. From then on, the forest cover increased to the current level of 30%, with periods of forest growth alternating with periods of decline. This fluctuation can be interpreted as the response to technological improvements in land use, to profound socio-economic changes or, more generally, as a mirror of culture (Mantel 1990; Harrison 1992; Bork et al. 1998; Küster 1998; Verheyen et al. 1999).

One of the first turning points came in the 14th century, as the decimation of the population caused by the plague affected broad stretches of land across Europe. The forests returned to places where the previous intensity of land use could no longer be maintained. The same occurred some 300 years later following the Thirty Years' War. After phases of forest overuse and destruction, systematic forestry since about the end of the 18th century has led to the expansion of forests to the current level, though the ecological character of the forests was changed significantly (Ellenberg 1988).

Today in large parts of Europe we are experiencing further, profound forces of change. First, supranational agricultural policies are leading to a decline in agricultural land use. The establishment of forests is one possible development approach on this land. A second process of change is the focus of this paper. It involves the structural changes in the old industrial regions of Europe and also North America. There, on sites where urban-

Kowarik I, Körner S (eds) Wild Urban Woodlands.

industrial uses since industrialization in the 19th century have created conditions not conducive to life that covered whole areas, today natural resettlement processes are revitalizing the areas as they grow into forests.

The emergence of new woodlands on profoundly changed man-made sites is, however, not a new phenomenon, if we consider, for example, succession on ancient ruins (Celesti Grapow and Blasi 2003). Small woodlands emerged as well on the fields of rubble in many destroyed European cities after World War II (Kreh 1955; Kohler and Sukopp 1964), though these woodlands were mostly razed again as cities became more built-up.

There are a few characteristics which suggest that the reforestation of urban-industrial areas should be recognized as a new type of process – a process that results as well in new kinds of demands related to work with new woodland areas. These characteristics include:

- The *extent of the reforestation processes*. Thousands of hectares of industrial land are involved, which were previously subjected to intensive use by iron and steel industries, and by mining and coal-working concerns with the slagheaps and rail yards that go along with these. Without intervention, in a few decades large complexes will arise consisting of the woodlands and the remains of the previous industrial use. At the beginning of the 1990s, in the German area of the Ruhr, more than 8,000 ha had already been abandoned (Tara and Zimmerman 1997). A multitude of these spaces will not be used for the foreseeable future.
- The *ecological configuration* of the urban woodlands. Profound changes to the sites, excavations, deposition of man-made substrates and also the influence of the surrounding city lead to habitats that deviate greatly from the expected character of the forest in terms of their communities and the related processes (Rebele and Dettmar 1996).
- The *spatial location* of the new woodlands. They are no longer found – as was once usually true – far outside the reach of most people, on the periphery of the cultivated landscape, where nature returned slowly and almost unnoticed. Far more often now, new woodlands emerge directly in the center of areas that formerly were highly economically active and therefore, within immediate reach of the public.

The need for a conceptual framework

Based solely on their location in the center of urban agglomerations, new urban-industrial woodlands could provide substantial social functions for the nearby residents and fulfill significant ecological functions. There are, however, obstacles in the way. The acceptance of the new woodlands by the local public, is, to put it carefully, uncertain. While "the forest" in certain areas is often held in a deep, symbolically rooted high regard (e.g. Harrison 1992), this is not the case with urban-industrial woodlands. With these, people are directly experiencing how an economic structure, which once functioned and from which they secured their existence, is being overwhelmed by elements of nature in a surprisingly short time. The decline of the former economic structure is therefore blatant. It is clear that the perception of the new, post-industrial nature is damaged by the stigma of the painful social changes that made such nature possible.

We know, from many studies, that the assignment of terms like "nature" and "wilderness" to concrete parts of the landscape varies a great deal, for example, between residents of urban and rural areas (Lutz et al. 1999; Bauer 2005), and depends, not least of all, on the ecological knowledge of the observer. So biologists are less inclined than others to describe urban green spaces as "nature" or "wilderness." In contrast, the tendency among residents to attribute a natural character to horticultural green spaces is greater (e.g. Jorgensen et al. 2005). In the classification of new urban-industrial woodlands, the discrepancies are even larger; the consensus about potential nature classifications is even more ambiguous, as the Leipzig study of Rink (2005), among others, demonstrates.

What shapes the character of the new urban-industrial woodlands is, briefly, the sharp contrast between a cultural layer of rubble, ruins and rusted iron and a natural layer that grows untamed and often surprisingly quickly. In the context of the former, one could diagnose the new woodlands as completely artificial, in the context of the latter as having a special wilderness character. Both, however, clearly belong together and create, from the characteristic double nature of urban-industrial woodlands, a product that is equally natural and artificial. This bipolarity does not, however, correspond to the general image of nature. Rather it brings about confusion and insecurity in the general classification as well as in the scientific classification of these spaces.

Consequently there are different classifications of urban-industrial woodlands as artificial or natural, as technologically shaped or as wilderness. It is clear that attached to such starkly different classifications will be significantly divergent assignments of value that will then become starting points for opposing concepts for development. Opportunities that would offer the potential social and ecological functions of the new woodlands for the development of the surrounding urban area may then be overlooked.

The perception of and the value given to urban-industrial woodlands is revealed not least through terminology, with whose help those value judgments are further conveyed. The main goal of this chapter is the development of a conceptual framework for classifying urban-industrial woodlands, one which allows their unusual features in comparison to other existing urban forest types, first, to be clearly discerned, and second, to be made semantically clear. Both are important prerequisites in order to be able to appropriately acknowledge the role of different types of urban woodlands and to be better able to make use of their social and ecological potential for further development of urban-industrial landscapes.

In the following sections, urban woodlands will be classified from different perspectives.

- First, types of urban and non-urban woodlands will be differentiated through a spatial approach.
- Then, how the influences of the urban surroundings ("the urban impacts") change the ecological characteristics of conventional woodlands will be described.
- Through a historical perspective, four types of forests will be differentiated in view of their emergence and use history and thereby a few characteristics of urban-industrial woodlands will be determined.
- Finally, how the nature and wilderness character of different forest types can be estimated by expanding the traditional classifications of naturalness will be presented. For this, non-scientific as well as scientific approaches will be used.

A spatial approach: urban and non-urban woodlands

Various types of woodlands may first be differentiated based on their location relative to urban areas. By this spatial approach, urban woodlands can be differentiated from peri- and non-urban woodlands (Table 1).

- Urban woodlands may be completely surrounded by developed areas and therefore be forest islands within the city. Most of the time, however, they lie on the city's fringe and have direct contact with developed urban areas on one side and with the open landscape on the other side.
- Peri-urban woodlands lie in the vicinity of the city and are deeply imbedded in the peri-urban cultural landscape. Most such peri-urban cultural landscapes were previously shaped by agriculture or village life. Today, however, they are mostly subject to increasing suburbanization.
- Non-urban woodlands, in contrast, lie far outside the urban impact area and are mostly interwoven with elements of the traditional cultural landscape.

This spatial organization generally corresponds to the different functions of the woodlands. With increasing proximity to the city, the accessibility of the woodlands for recreation-seeking urban residents grows and thereby so does the woodlands' social function (e.g. Roovers et al. 2002). To the same extent, the ecological function of the woodlands in terms of traditional wood production and hunting yield decreases. It would be rash, however, to start from the assumption of a general replacement of economic functions with social ones in urban woodlands. It is far more true that social and ecological functions ("ecosystem services") can also be given value (Shaw and Bible 1996; Costanza et al. 1997; Tyrvainen and Väänänen 1998; Tyrvainen 2001).

Table 1. Spatial differentiation of urban, peri-urban, and non-urban woodlands according to their location relative to urban areas. In general, with changes in the location of the woods, their social and production functions change as does the level of influence of urban impacts

Forest type	Sub-type /Description	Spatial characteristics	Function social	Function production	Urban impacts
Urban woodlands	Woodlands within urban areas	Isolated in built-up areas			
	Woodlands on the urban fringe	Between built-up areas and the open landscape			
Peri-urban woodlands	Woodlands in the vicinity of urban areas	Part of the open (cultural) landscape close to urban areas			
Non-urban woodlands	Woodlands far from urban areas	Part of the open (near-natural) landscape far from urban areas			

An ecological perspective: urban impacts change traditional woodlands

Numerous studies have shown that urban conditions have significant impacts on climate, soils, and biodiversity patterns. Such changes are effected with decreasing intensity along urban–rural gradients (Kunick 1982;

McDonnell and Picket 1990; McKinney 2002; Baxter et al. 2002; Niemela et al. 2002; Venn et al. 2003). Urban conditions have effects as well on the types of woodlands, which are ordered in Table 1 along a urban-rural gradient. In the following, a few of such changes will be described. The next section will describe how, on urban-industrial sites, new types of woodlands may emerge that differ profoundly from the original ones.

Habitat fragmentation

Urban uses often lead to the fragmentation of woodlands and to stark divisions within stands through intended and unintended paths and horseback-riding trails as well as roads. In this way, small woodland patches with high edge-to-interior ratios are created. Forest fragmentation generally enhances pioneer species or non-native species that respond well to an increased availability of light (Brothers and Spingarn 1992; Godefroid and Koedam 2003a). Fragmentation by roads supports species that are dispersed by vehicles (Parendes et al. 2000; Ebrecht and Schmidt 2003). However, typical forest plant species, including rare species, may also occur along forest edges (Godefroid and Koedam 2003b). Bird species are often negatively affected by decreasing patch size (e.g. Mortberg 2001). An increasing density of paths can negatively affect the establishment of saplings (Lehvavirta and Rita 2002), but the regeneration of trees is generally not threatened in urban woodlands. Due to varying responses of tree species to fragmentation, changes in the species composition, however, may occur. The ground layer is most susceptible to trampling, which leads to a decrease in the vegetation cover (Malmivaara et al. 2002).

Deposition of materials

Peri-urban woodlands are especially affected by wet or dry deposition as well as that resulting from recreational activities. The spread of nitrophilous species is interpreted to be a result of nitrogen deposition (Kowarik and Sukopp 1984). Müller et al. (1978) correlated the presence of high-N indicator species (Ellenberg's indicator species) with the use frequency of forest paths by recreationalists. The distribution of macrofungi in pine forests can also be related to an urban pollution gradient (Tarvainen et al. 2003). Deposition from urban-industrial sources can also clearly balance out the limiting effects of the original conditions of the site. In this way, within peri-urban pine–oak forests, trees that are more demanding are spreading across what were originally nutrient-poor sandy sites (e.g. *Acer platanoides*, *Acer pseudoplatanus*; Fischer 1975; Sachse et al. 1990). The deposition of industrial materials with a high pH suits lime-loving species, especially in areas with naturally acidic soils. The establishment of the North American *Mahonia aquifolia*, for example, is promoted by an increase in the pH values from fly ash (Auge 1997).

Exchange of species

The spatial interweaving of urban woodlands and developed areas favors the exchange of species in both directions. Numerous cultivated plants disperse themselves as escapees from gardens and parks into neighboring woodlands or are carried into such areas as garden waste (Hodkinson and Thompson 1997). Urban woodlands are generally rich in non-native species especially along their edge areas (Asmus 1981; Moran 1984; Walther 1999). In the other direction, attractive forest plants are usually transplanted into urban gardens (Kosmale 1981) where they survive, but generally are not able to spread, in contrast to the escapees. Forest species with the capacity for long-distance dispersal, for example ferns, orchids and flying ground beetles, may also colonize urban-industrial sites (Dickson 1989; Keil et al. 2002; Weiss et al. 2005). A few highly mobile animal species profit as well from attractive food offerings in bordering gardens and parks. In this way, in a few peri-urban woodlands of Germany, North American raccoons have reached population sizes comparable to those in their home territories (Hohmann and Bartussek 2001). In Berlin, wild boars use gardens and parks near woodlands to search for food; foxes are found in city centers as well.

Impacts on forest species

Recreation use influences the populations of forest species in very different ways. For peri-urban woodlands with good access, frequent disturbance by recreationalists and above all, by dogs, is a given. Predatory animals such as dogs and house cats can have negative effects on bird populations (Marzluff 2001). They can also cause game to retreat farther into the forest, which can, in turn, have positive effects on plants that are preferentially browsed. The yew (*Taxus baccata*) has become established in parts of Berlin's Grunewald and can grow large there, something that is usually hindered by browsing (Seidling 1999). Dogs can also disperse forest plants, whose seeds are transported on the dogs' coats (Graae 2002).

In peri-urban woods, the silvicultural management strategies are often aimed at an aesthetically attractive forest structure. The presence of old and mature trees and stands is regarded as an important feature in the aesthetic perception of forests (Ode and Fry 2002). Through promotion of old individual trees and increase in the percentage of dead wood, specific, mostly threatened, animal species will be encouraged, namely insects that require decomposing wood as a habitat or sustenance basis and tree-cavity residents such as bats and cavity nesters.

A historical approach: from pristine to urban-industrial woodlands

Independent of their location relative to urban agglomerations, peri-urban as well as non-urban woodlands can vary tremendously in their origins and development history. In addition to the spatial dimension illustrated in Table 1, the urban woodland matrix is also determined by its origins. In this second dimension, four woodland types can be differentiated (Table 2). They differ fundamentally in regard to habitat continuity and to the agency of natural versus cultural mechanisms. It should be said, however, that the illustration in Table 2 presents idealized types. In practice, there are numerous points of overlap, just as there are narrow spatial overlaps and interweavings between the various types of woodlands.

For our understanding and also for developing urban woodlands, the conditions of their origins are of particular importance. A number of qualitative features result from this history that ensure certain social and ecological functions. Such features include, for example, the age, the species composition, the stand structure of the woodlands, the existence of certain natural elements (e.g. cliffs, bodies of water) or artifacts and also infrastructure elements such as parking lots, walking and riding paths, picnic areas, etc.

Remnants of pristine forests

Remnants of pristine forests are usually understood to be non-urban forests, far outside the impact area of cities. They can, however, occur as peri-urban woodlands and even as isolated urban woodlands in the centers of cities. An example of the latter is the oak–hemlock forest that today lies in the middle of the Bronx as a part of the New York Botanical Garden; it represents a remnant of the vast forests that once covered the East Coast of North America (McDonnell and Rudnicky 1989).

In the middle of urban areas, and farther away as well, such remnants of pristine forests are not usually free from human influences. As a rule they are affected by anthropogenic depositions, by at least minor historic or contemporary forestry uses or by recreation activities. Though a few cultural influences may have an effect on them, it is fundamentally the natural processes of the site and stand dynamics which shape their species composition and structure. An additional important feature of remnants of pristine forests is their high habitat continuity. This makes possible, among other things, the presence of less mobile animal and plant species that are described as specialists of old forest sites (Peterken 1974, 1994; Verheyen et al. 2003; Wulf 2003).

Table 2. Historical differentiation of woodlands according to the history of their development and use. The four types of woodlands vary conspicuously in terms of the agency of natural versus cultural mechanisms and in their habitat continuity. The forest types are further classified according to the "four natures approach," see text for explanation). H – Habitat continuity; N – Agency of natural mechanisms; C – Agency of cultural mechanisms

Types of forests	Types of ecosystems	Types of "nature"	H	N	C
Remnants of pristine forests	Pristine ecosystems	Nature 1: "old wilderness"			
Forests strongly characterized by silviculture	Ecosystems shaped by silvi-/agriculture	Nature 2: "traditional cultural landscape"			
Planted tree stands in green spaces	Ecosystems established by urban greening	Nature 3: "functional greening"			
Woodland succession on urban-industrial sites	Ecosystems evolved on urban-industrial sites	Nature 4: "new wilderness"			

Woodlands as a part of the traditional cultural landscape

Virgin, natural woodlands were transformed into elements of the traditional cultural landscape when their structure and species composition were heavily influenced by historical or modern silvicultural uses (e.g. woodland pastures, reforestation partly with non-native species; Pott and Hüppe 1991; Zerbe 2004). The habitat continuity of such woodlands can still be quite high when it is a question of transformed, pristine forests. The continuity on reforested former agricultural land, however, is comparably low. For parts of Europe, more recent reforestations have been differentiated from old-growth forests through landscape historical methods (Peterken and Game 1984; Peterken 1994; Wulf and Kelm 1994; Verheyen et al. 2003). According to these studies, the forest types differ in their species richness and in the presence of indicator species of old-growth forests.

The development of two peri-urban forests in Switzerland and in Germany illustrates possible overlaps between culturally determined and pristine natural forests. The Sihlwald has been fundamentally shaped by the long-term procurement of wood for the firewood needs of the city of Zürich (Broggi 1992). The Saarkohlenwald near Saarbrücken has also been shaped by traditional forestry uses – and additionally through small-scale mining activities (Rösler 2004). The goal in both cases of allowing a peri-urban wilderness to emerge by abandoning former uses will lead over the course of time to a convergence of the stands with the species composition and structure of a pristine woodland.

Woodlands that result from urban greening

Stands of trees that present forest characteristics can develop from large-scale planting of trees in park areas, but also from street trees and restoration plantings. Because such stands are determined based on functional goals and are often designed and maintained from artistic-aesthetic perspectives, woodlands resulting from urban greening are heavily culturally influenced.

An additional feature that differentiates this type of woodland from the two previous types is the high species diversity and the large role of non-native species in urban greenery (Ringenberg 1994; Freedman et al. 1996; Jim and Liu 2001; Kowarik 2005). The native species also often deviate genetically from regionally native plants because they are dispersed over the supraregional nursery network, which in large part imports its seed material from other, off far-removed areas. The hazel (*Corylus avellana*) planted in Germany, for example, usually originates in southern Italy or Turkey (Spethmann 1995).

Despite their cultural basis, woodlands of urban greenery can also be significantly shaped by natural processes. These include the natural regeneration of tree species or the establishment of wild plants within planted stands as well as the "export" of planted species that escape from cultivation. In this way, the probability of a species becoming naturalized is a direct function of the number of plantings (Rejmanek 2000; Mulvaney 2001). Prominent examples of this are the substantial dispersion of maple species (*Acer platanoides*, *A. pseudoplatanus*) from horticultural plantings (Sachse 1989; Webb and Kaunzinger 1993; Preston et al. 2002).

The interaction of cultural (e.g. cultivation) and natural processes (e.g. the establishment and spread of cultivated species) often leads to park-specific communities in urban green spaces. This is clear above all in old historic gardens, in which a few plant species that were cultivated in earlier Baroque or landscape gardens have survived and have, in some cases, established large populations. These can be interpreted as cultural remnants or as indicators of former horticulture (Bakker and Boeve 1985; Kowarik 1998).

In addition to new plantings, woodlands can also emerge within the context of urban greening through the transformation of pristine forests or woodlands used for forestry purposes. In this way, the Tiergarten park in the center of Berlin developed from an old-growth forest that experienced a period as a Baroque park in the 18th century and was redesigned in the 19th century as a landscape park (Wendland 1979). Above all, since the end of the 19th century, peri-urban parts of woodlands have been artistically shaped in order to improve their social function (e.g. the Eilenriede in Hannover; Hennebo 1971).

The example of the new development Birchwood in England illustrates the opposite case: the convergence of planted stands toward a near-natural model. In the planning concept, the existing remnants and newly planted woodlands were integrated as space-creating elements. The new plantings were integrated with the existing old stands through a "naturalistic style" (Jorgensen et al. 2005).

Urban-industrial woodlands: a new woodland type emerges on specific urban sites

As post-industrial uses are often not possible in the areas affected by structural change, the abandonment of sites is a signal for the development of a new type of urban-industrial forest by natural colonization processes. In the German Ruhr, this new forest type is widely dispersed across sites of the iron and steel industries and across mining areas (Dettmar 1992; Weiss et al. 2005). New urban woodlands can emerge on other types of sites as well, for example, on old, less-maintained cemeteries (e.g. Zisenis 1996), on the rubble of former buildings (Kohler and Sukopp 1964; Kowarik 1992a; Fig. 1–5) and on rail yards that have fallen into neglect (e.g. Reidl and Dettmar 1993; Kowarik and Langer 1994; Burckhardt et al. 2003).

A few peri-urban woodlands can be equivalent to urban-industrial woodlands when industrial uses away from cities lead to significant site

Table 3. Urban-industrial woodlands result both from cultural and natural processes

Cultural processes	Natural processes
Abiotic conditions	
• Earlier uses have significantly altered substrates or soils have been created entirely anew	• Soil formation after excavation or deposition of substrates with the accompanying physical-chemical processes (e.g. decalcification, alkalinization, enrichment of organic matter)
• Structural relics of earlier uses may remain in their entirety or in part	• Corrosion or erosion of structures
Biotic conditions	
• Relics of earlier horticultural plantings may still be present	• Proliferation and establishment of formerly cultivated plant species
• Previous uses may have unintentionally introduced non-native plant or animal species	• Proliferation and establishment of unintentionally introduced species
• The culturally determined species pool of the urban surroundings acts as source for the colonization of derelict sites; significant role of non-native species as ornamentals	• Invasion by escapees from cultivation, establishment and spread of new populations; significant role of non-native species as escapees
	• Colonization by components of pristine vegetation and by animal species (mostly native species with long-distance dispersal)
	• Growth and decline of populations in the course of succession and natural stand dynamics
• After abandonment of earlier uses, social activities may change the vegetation and site dynamics and lead to new cultural patterns	• Adaptation of the vegetation and site dynamics to current uses

changes. Post-mining landscapes are a good example (Tischew and Lorenz 2005). Urban-industrial woodlands generally can be distinguished through fundamental characteristics from the other types described earlier.

Habitat continuity

In contrast to the other woodland types, the habitat continuity of urban-industrial woodlands is generally very low. Forest stands, however, can develop within two to three decades. These merely decades-old stands predominate on abandoned areas in former industrial regions of central Europe. As is expected with such a short development time, the soils are not yet fully developed with the result that the vegetation development is heavily dependent on the characteristics of the anthropogenic substrate. In certain situations, however, woodlands of profoundly changed urban-industrial sites may be distinctly older, e.g. those on ancient ruins, fortress grounds, or old mine slagheaps (Dickson 1989; Jochimsen 1991; Caneva et al. 2003; Celesti Grapow and Blasi 2003).

Agency of natural mechanisms

The leeway for the agency of natural mechanisms is significantly greater in urban-industrial woodlands than in woodlands used for forestry or in those resulting from urban greening. The latter two are usually heavily influenced by the initial plantings, and by use and maintenance. On urban-industrial abandoned areas, in contrast, the ecosystem dynamics are mainly influenced by natural processes through population dynamics, succession and soil formation (Table 3; right column). In this way – depending on the extent of the changes caused by earlier uses – primary or secondary succession as well as intermediate degrees can be induced (Rebele 2003). In cases where current social uses are limited or do not exist due to limited accessibility, the development of urban-industrial woodlands is fundamentally determined by natural processes just as it is in pristine forests. From a few studies of succession, we know that dispersal, establishment, and extinction processes can effect substantial changes to the population composition in time frames of a few years to a few decades, for example the replacement of pioneer species with forest species (Platen and Kowarik 1995; Weiss et al. 2005).

Agency of cultural mechanisms

In contrast to pristine forests, urban-industrial woodlands are heavily culturally influenced. This is true of abiotic conditions as well as biotic (Table 3; left column). Through removal and deposition of a variety of anthropogenic materials, the spatial and functional contact with the original soil and its hydrology is disturbed. Woodland development on many urban-industrial sites is based, therefore, on a new type of abiotic site configuration that may deviate significantly from the original conditions. The soil development of coal-mining slagheaps is influenced by particularly low pH values, while soils that develop on building rubble may present very high pH values (Rebele and Dettmar 1996). Extensive excavations can lead to nutrient-poor substrates upon which soil formation then proceeds within the framework of primary succession (e.g. Tischew and Lorenz 2005).

Previous uses can, however, continue to influence the site or vegetation development or may remain present in the form of structural ruins. As these vary in form, extent, and frequency, clear spatial patterns often emerge in the landscape morphology or the vegetation, which make the earlier uses readable. Through the effects of natural processes, these relics, however, lose their distinctness over time. The development of urban-industrial woodlands is also shaped indirectly through the urban surroundings. This has an effect on the species pool that provides a source for the colonization of the site.

Species pool

A fundamental feature of the urban species pool is the prevalence of non-native species that were introduced accidentally to urban habitats or that escaped from cultivation (Kowarik 1995). As a consequence, non-native species play a large role in reforestation processes on urban abandoned areas, but are significantly less important on peri-urban sites, for example, during succession on surface mines (Prach and Pysek 1994; Tischew and Lorenz 2005). Table 4 illustrates the significance of non-native species with the example of woody species that are dispersed in different habitats of Berlin.

The dispersal processes that lead to the colonization of abandoned areas are determined culturally as the choice of species planted in the surroundings of the site that then serve as sources for further dispersal was horticulturally determined. In Berlin's inner-city abandoned areas this has led to a heterogeneous pattern in the distribution of stand-building tree species. In the colonization processes of peri- and non-urban abandoned areas, it is usually native early-successional trees such as European white birch (*Betula pendula*) and European aspen (*Populus tremula*) that dominate. These are found in inner city areas as well. When competitive non-native species have been cultivated in the vicinity, however, their dispersion can lead to new types of urban-industrial woodlands, as the dominance of the North American *Robinia pseudoacacia* in the former *Diplomatenviertel* (Diplomatic Quarter) of Berlin demonstrates (Kowarik 1992a; Fig. 1). In the Ruhr, the influence of urban dispersion sources is revealed in the widespread establishment of the ornamental *Buddleja davidii* in early successional woodlands that are dominated by native birches (Dettmar 1992).

A few native forest species may colonize even isolated urban sites by natural means, such as some ferns (Keil et al. 2002). When forest remnants lie in the vicinity of the succession area, however, typical forest species can establish themselves over time (Tischew and Lorenz 2005). Most forest species, however, can not overcome the distance between remnants of pristine forests and urban-industrial sites. In city centers, forest development on abandoned areas is thus mostly determined by species that migrate from the urban surroundings.

When non-native species colonize derelict land, the connection to horticultural dispersion sources in the surrounding areas and also to the relics of earlier plantings on the site is often very clearly readable as a culturally based process. Less obvious is the parallel case of native species. Because native species are also mostly introduced from other areas via horticultural distribution and partly demonstrate features of domestication, they form a component of the species pool that deviates genetically from the original species composition (Kowarik 2005). Native escapees from cultivation, like non-native ones, can therefore change the species composition or strongly influence the stand development of urban-industrial woodlands. Both cases lead to altered biodiversity patterns in comparison to the other woodland types described in Table 2.

Table 4. Woody species (trees, shrubs, woody vines) escaped from cultivation and occurring in different urban ecosystems of Berlin. All non-native species are escapees. For native species, the number of species descended from natural populations or from cultivated populations is unknown (after Kowarik 1992a). Column 1 corresponds to the "four natures approach" (see text for explanation)

Type of nature	Type of ecosystem	No of species	% native	% non-native
Nature 1	Wetlands in the urban fringe (remnants of pristine vegetation)	40	88	12
Nature 1, 2	Woodlands at the urban fringe (remnants of pristine and managed forests)	141	43	57
Nature 3	Urban parks	171	30	70
Nature 3 (4)	Residential areas (developed areas including gardens)	155	34	66
Nature 4	Woodlands and other succession stages in abandoned urban-industrial areas	173	33	66

Fig. 1–5. Cultural layers and natural processes in urban woodlands. The Diplomatenviertel in the center of Berlin was heavily damaged during World War II and was then cleared, leaving only a few ruins. The woodlands that mainly resulted from natural colonization processes illustrate, directly and indirectly, the influences of previous uses, particularly earlier horticulture. The dead black locust (*Robinia pseudoacacia*) in the upper left came from a former embassy garden. It was a dispersion source for the expansive wild woodlands, now dominated by *Robinia*. A few relics of old plantings point directly to the 19th-century garden history of the site (*Fagus sylvatica*, *Catalpa bignonioides*). In the herbaceous vegetation, another North American species, *Parietaria pensylvanica*, occurs, whose German distribution was limited to Berlin for much of the post-War period. According to Sukopp and Scholz (1964), the seeds of these species that escaped from the Botanical Garden may have been widely dispersed throughout the city by the turbulence caused by bombing raids. Through informal uses, open areas are still maintained. The picture below shows the changes in the soil morphology from jumping areas for mountain bikers. Here pioneer species, such as the eastern Mediterranean *Chenopodium botrys*, that were typical of the new rubble fields of the early post-War period can survive. The naturalness of the woodlands, from a retrospective perspective is low. The naturalness of the site from the second, prospective approach is, in contrast, relatively high, because the leeway for natural processes in these urban woodlands is large.

Are urban-industrial woodlands natural?

The question of the naturalness of urban woodland types leads, undeniably, to starkly varying answers. The differences have to do with the different histories of the stands (Table 2), and also with differences in perceptions of naturalness, which vary dramatically across different geographic reference areas and between different social groups (e.g. Henderson 1992; Ewert 1998; Lutz et al. 1999; Bauer 2005). How the characteristic double nature of urban-industrial woodlands as a product of both nature and culture should therefore be classified remains an open question because traditional scientific approaches to the classification of ecosystem types are usually oriented toward pristine ecosystems. A precise answer is, however, necessary, when, for example, a "near-natural" development of forest stands is demanded in public projects. What is meant by this, however, remains seemingly unspecified.

In this section it will be demonstrated that the particular characteristics of the woodland types shown in Table 2 can also be described in terms of their "naturalness" and "wildness." This would seem to be an important prerequisite in order to make the best possible use of the varying social and ecological potentials of the different woodland types in the development of urban green spaces. Because these potentials are to be made transparent for different interested parties, for local residents as well as for stakeholders and scientists, a differentiation is made below on two levels.

Scientific approaches to naturalness

For about one hundred years, numerous approaches for scientifically classifying the naturalness of vegetation types and ecosystems have been developed. They share the fact that they evaluate varying degrees of naturalness or, reciprocally, they evaluate the extent of human influence. They share the further characteristic that a defined reference point of maximum naturalness is often missing or at least remains undetermined. Two perspectives on naturalness can be fundamentally differentiated: naturalness from a retrospective or a prospective perspective (Kowarik 1988, 1999). Table 5 illustrates this concept through the assignment of existing classification approaches to both perspectives. The deciding factor is the reference point. It may lie in the past or it may be adopted from the present or the future.

Naturalness from a retrospective perspective

The retrospectively (i.e. vegetation historically) oriented perspective analyzes the extent to which a current woodland corresponds to an earlier stand that grew in the same region and whose structure, species composition and site factors were not influenced by human activities. The point of reference is therefore, pristine vegetation uninfluenced by humans. Based on the cultural history of the relevant area, the reference period may lie decades or a few millennia in the past. Such a historically based comparison is fundamentally made possible with the help of vegetation-history knowledge. Different scales have been developed to estimate the different grades of "historic" naturalness; the scale of Ellenberg (1963), with seven degrees from "untouched" to "artificial," is the most widely used (for more details see Kowarik 1999).

In the retrospective perspective of naturalness, remnants of pristine woodlands are most natural and woodlands used for forestry are at least semi-natural. Should the silvicultural use of these woodlands, such as the Sihlwald near Zürich and the Saarkohlenwald near Saarbrücken, be reduced, forest development will lead to a convergence to pristine vegetation and therefore back to a "retrospectively" natural stand.

In contrast, horticulturally planted and maintained woodlands are, from a vegetation history perspective, artificial, as the species composition, stand structure and frequently the site conditions deviate substantially from pristine woodland types. How are new types of urban-industrial woodlands assessed? In the traditional, retrospectively oriented analysis, these woodlands, based on their cultural shaping, must be considered "artificial" when, under anthropogenic site configurations, stands grow that are dominated by species from the urban species pool rather than the historic species pool. This has the consequence that urban-industrial woodlands are assessed, in the retrospective approach, as just as artificial as horticulturally shaped stands, even when they are profoundly shaped by the natural processes described above.

In summary this means that from the retrospective perspective the development *back* to nearly natural or natural woodlands can be analyzed well. With the evaluation of *new* development of "wild" urban-industrial woodlands, however, the traditional concept of naturalness oriented toward historical comparisons runs aground. The retrospective approach does not recognize the clear wilderness character of such stands and therefore can not be used to differentiate these woodlands from other stands and allow for a differentiated evaluation.

Naturalness from a prospective perspective

The retrospective approaches, which evaluate the naturalness of areas by comparing them with conditions that haven't been influenced by humans, have a long tradition. This is due to the fact that the specific natural development of urban-industrial sites was recognized in ecological research relatively late. There are, however, a few authors, little-recognized today, that early on suggested a second perspective for evaluating naturalness. Bernatsky (1904) wrote that anthropogenically shaped vegetation types could regenerate themselves to an "*Urformation*" (a virginal or primeval condition), but that a development to a "natural condition" was also possible when the earlier cultural influences continued to have an effect. The forestry scientist von Hornstein (1950, 1954) demonstrated that a new type of forest, not previously existing, could grow on severely altered sites, namely a "tertiary type."

As a continuation of such an approach, a second type of naturalness can be identified beyond a historic naturalness. For this, the reference point is not an original condition of a natural landscape, but rather a condition is defined based on the current site potential and the greatest possible degree of self-regulation (Kowarik 1988, 1999). From this perspective, therefore, the natural capacity for *process* is the central point, not a particular, retrospectively determined and often idealized, *picture* of nature. This perspective can be described as contemporaneous, because it begins from the existing potential of the site without regard for whether this has been exclusively naturally or anthropogenically shaped. It can also be described as prospective in that the naturalness of a site is evaluated according to how far, based on current uses for example, the stand is from a future condition of self-regulation without direct human influence. The hemeroby system is used to evaluate the degree of this second kind of naturalness; this system was created by Jalas (1955) and more precisely defined by Kowarik (1988) with regard to its contemporaneous orientation (Table 5).

Remnants of pristine woodlands are evaluated in this second, prospective view of naturalness as just as "natural" as they are in the retrospective view because they are determined by self-regulation of abiotic and biotic processes to the fullest extent possible. Distinctions are possible, however, when non-native species such as black cherry (*Prunus serotina*) can become established in the stands in the long term. From the retrospective view, this would stand in the way of a ondition; this is not true in the prospective approach (for a description of the scientific conflicts that arise from this, in setting goals for national parks and in process conservation, see Kowarik 2003, Körner 2005).

In the evaluation of the horticulturally shaped stands of urban green spaces, diverging assessments are probable. Woodland stands of horticulturally introduced non-native plants are always considered "artificial" in the retrospective perspective, even when other species have become established and the original planted species have propagated and spread through

Table 5. Concepts for classifying ecosystem or vegetation types according to their naturalness from a retrospective and a prospective perspective (adapted from Kowarik 1988, 1999)

Approach to defining naturalness	Reference condition	Allows for the integration of irreversible human alterations to the ecosystem	Advantages (+) and disadvantages (-) for nature conservation evaluations	Assignment of existing naturalness classifications
Retrospective (historical naturalness)	Pristine ecosystems, without cultural impacts from humans	No	(+) Identification of remnants of pristine ecosystems possible; differentiated assessment of changes to the original character possible (-) Differentiated assessments after substantial ecosystem changes not possible; focused on a historic picture of nature that often can not practically be reproduced	Naturalness sensu Westhoff 1949, 1951; von Hornstein 1950, 1954; Ellenberg 1963; Seibert 1980. Synanthropization sensu Falinski 1966, 1986. Naturité sensu Géhu and Géhu 1979
Prospective (contemporaneous naturalness = hemeroby)	Condition of an ecosystem that is shaped by self-regulation	Yes	(+) Differentiated assessments after substantial ecosystem changes possible; open to assessing new developments of nature, subsequent to human impacts or "global changes" (-) Naturalness is assessed ahistorically; certain values of the remnants of original ecosystems are not recognized	Hemeroby sensu Kowarik 1988 (based on a more precise definition of Jalas 1955; Sukopp 1969, 1972). Anthropopressure sensu Olaczec 1982. Artificialization sensu Long 1974. Naturalness sensu Miyawaki and Fujiwara 1975; Schlüter 1984, 1992

natural regeneration processes and have, over the course of time, replaced the planted individuals. In the prospective perspective, the naturalness degree of such stands would increase during the course of such processes.

Clearly, the two perspectives on naturalness will result in diverging assessments for woodlands resulting from succession on severely altered urban-industrial sites. Profound changes to the sites as well as the dominance of species from the urban species pool lead to significant deviations from historic vegetation types. A convergent development to pristine vegetation types is definitely possible, for example, on mining slagheaps, where mixed-hardwood forests may grow on acid soils. Studies in Berlin have made clear that non-native species such as *Robinia pseudoacacia* may result in divergent succession trends even in the long term (Kowarik 2003). For this second, prospective perspective of naturalness, however, the origin of the species is irrelevant as is an irreversible, anthropogenic transformation of the morphology, soils and moisture regime. Urban-industrial woodlands are assessed as natural to the extent that they are free from direct, reversible human impacts and that they have reached a high level of self-regulation, for example, in the population dynamics of the plant species. This is, in principle, characteristic of pristine natural woodlands as well.

On urban-industrial sites, the prospective approach to classifying naturalness allows, for instance, for the differentiation between planted woodlands (e.g. for ornamental or reclamation purposes) and woodlands resulting from natural colonization processes. It further makes possible a differentiation between woodlands that are shaped by human uses to different degrees. This may lead, in certain urban agglomerations, to the surprising view that, of the ecosystems in the impact area of cities, it is the new types of urban-industrial woodlands that reach the highest level of naturalness when other sites are very heavily influenced by urban, forestry, or agricultural uses.

The "four natures approach": A non-scientific approach to urban nature

In contrast to the two natural-science perspectives presented above, the understanding of "naturalness" and "wilderness" for the general public, but also for local stakeholders, is generally broader, but also individually very different, as studies by Rink (2005) and Bauer (2005) show. This runs the risk that significant qualitative differences between the different types of urban forests may not be appropriately realized and included in decision-making processes, and that nature and wilderness attributes will be used with less comprehensibility. As a complement to the two science-based approaches described above for assessing the naturalness of ecosystems, fundamental differences between the different types of forests can be made generally understandable with a simple organizational approach.

The basic idea of the "four natures approach" (Kowarik 1991, 1992b) is to reduce the existing diversity of very different, culturally varying forms of nature that are found within the impact area of cities down to four types and thereby emphasize their respective "characters" (Table 2).

- "Nature of the first kind" is the "original" nature. This includes ecosystem types or landscapes that are remnants of pristine ecosystems or at least areas that still have a strong connection to pristine ecosystems despite certain uses or changes through urban impacts. These include, for example, old-growth forests, moors, some rivers and lakes, rock formations, etc.
- "Nature of the second kind" includes elements of the landscape that arose through traditional or modern agriculture and forestry practices, for example, meadows and pastures, crop fields, hedgerows, coppices, intensively managed forests, etc.
- "Nature of the third kind" comprises the greenery that has emerged through horticultural plantings, maintenance and upkeep. This includes, fundamentally, gardens and parks created during different eras of garden history, but also other urban greenery such as street trees or trees planted to define spaces in developments. This "third nature" was first recognized in the Renaissance (de Jong 1998).
- "Nature of the fourth kind" encompasses the natural development that occurs independently on typical urban-industrial sites, without horticultural planning or design. This starts with cracks in sidewalks or in colonization of walls and buildings as "artificial cliffs" and leads to growth in abandoned areas and to impressive urban-industrial woodlands.

While natures of the second and third kind are heavily culturally influenced, the first and fourth are more defined by the effects of natural processes. It is easiest to assign a wilderness character to these types (Table 2). In order to keep the fundamental differences between them transparent, one can speak of nature of the first kind as "old wilderness" and nature of the fourth kind as "new wilderness." "New wilderness" is also occasionally used when wilderness character returns to silviculturally influenced forests or after the initiation of natural processes during reclamation activities. After the decline in forestry uses, a new wilderness development does in fact take place. It is, however, not qualitatively new, but rather involves a convergence toward an "old" wilderness. The same occurs when restoration is undertaken with the goal of approaching original conditions. In contrast, a qualitatively new type of wilderness emerges on many urban-industrial sites that, for this reason, will be described as "new wilderness" in agreement with Dettmar and Ganser (1999).

The "four natures approach" pursues two main goals through its simple differentiation:

- The first goal is, in view of the overwhelming variety of concrete *manifestations* of nature, to make fundamental differences between *types* of nature more transparent for the general public. In this way, the particular character of each type can also be better acknowledged during the planning process. To this extent, the "four natures approach" means, on one hand, an abstraction of the existing diversity; on the other hand, it also represents a qualitative differentiation, in contrast to more sweeping categorizations such as "urban green" or "urban forests."
- The second goal is to convey, through a simple distinction between natures of the first, second, third and fourth kind, that a fundamental equivalence of values exists among the four different nature types. The original nature, Nature 1, which is identified as the "correct" nature from a scientific perspective through the application of the retrospective perspective of naturalness (see above) is therefore not automatically more valuable that the other manifestations of nature. An urban-industrial woodland can also be identified as especially valuable.

The advantage of the "four natures approach" can be summarized as the recognition of the existence of fundamentally different types of nature in urban-industrial landscapes (and beyond). It allows value to be assessed, in the context of nature conservation evaluations for example, not only in terms of traditional images of nature, such as former wilderness or traditional cultural landscapes; but also allows for a certain landscape park or even a certain form of an urban-industrial woodland to be recognized as exceptionally valuable.

The value of a certain area is, therefore, not determined beforehand according to its typological level (1 is better than 2, 2 is better than 3, etc.). Instead such an evaluation always results based on the manifestation of the type of nature at the object level. What is decisive here is the question of what manifestation or what development potential is concretely at hand and to what extent this corresponds to the particular planning or development goals.

Until now, at least in Germany, nature conservation goals have been oriented more toward the traditional image of nature (see Körner 2005). The "four natures approach" offers a perspective for new types of nature development that have been culturally influenced in various ways. It increases the chances that the social and ecological potentials of the different types of nature can be better used for the development of urban landscapes.

Conclusions

Wild urban woodlands resulting from natural succession on man-made sites have created a new component in the urban forest mix whose significance will grow in areas that are subjected to great structural transformation. These include many former industrial areas, but also, more generally, "shrinking cities." A particular feature of the new urban wilderness is its position in the middle of urban agglomerations. This represents a great potential to bridge, at least partially, the often lamented spatial separation between a large part of the general public and real, existing biodiversity. This distance has grown more acute as a consequence of urbanization. Because of the significant social and ecological functions of urban-industrial woodlands, strategies should be promoted that incorporate such spaces into the development of urban green spaces. A few conditions, however, are necessary first.

It is reasonable to reconsider the often one-sided focus of nature conservation on original nature. Until now, most conservation strategies in urban agglomerations have focused on (1) the preservation of remnants of pristine ecosystems, or (2) on restoring native species in managed or ruderal habitats (e.g. Kendle and Forbes 1997; McKinney 2002). The evolution of specific urban-industrial ecosystems, especially of a new type of wild urban woodland calls for an additional third way: the acceptance of natural processes that lead to a new kind of post-industrial ecosystem. This "nature of the fourth kind" as a "new" wilderness sharply diverges from t9he traditional "old" wilderness of our pristine ecosystem relics. However, how the conservation value of ecosystem development on urban-industrial sites should be evaluated usually remains uncertain (Harrison and Davies 2002).

In Britain, former mining slagheaps have long been seen as "special sites of scientific interest" with substantial conservation value (Kelcey 1975; Davis 1976; Box 1993). A very few projects in Germany, such as the Industriewald Ruhrgebiet (the Industrial Forest of the Ruhr) or Berlin's Südgelände demonstrate that urban-industrial nature can also be incorporated in the development of green spaces within developed areas (Weiss 2003; Kowarik et al. 2004; Kowarik and Langer 2005; Dettmar 2005).

We know from various studies, however, that residents often have reservations about wilderness. What is partly true of traditional wilderness areas (e.g. Durrant and Shumway 2004; Bauer 2005) is all the more true for the "new wilderness" of urban-industrial woodlands. It is therefore necessary to make this new type of nature more accessible; design measures, the incorporation of works of art and an intensified work with the public may be of help. In the development of urban-industrial woodlands, as is the case in historic parks (Kowarik et al. 1998), there exists a shared area of responsibility for disciplines that often work separately from one another. In addition to nature conservation and landscape architecture, historic preservation should also be incorporated when the cultural foundation of the new nature development within post-industrial landscapes has historic preservation value. Common historic roots can be drawn upon for this, as Körner (2005) shows with the example of nature conservation, historic preservation, landscape architecture and forest aesthetics.

Acknowledgements

For comments on the text, I would like to thank Stefan Körner and for the translation, Kelaine Vargas.

References

Asmus U (1981) Der Einfluß von Nutzungsänderung und Ziergärten auf die Florenzusammensetzung stadtnaher Forste in Erlangen. Ber Bayer Bot Ges 52:117–121

Auge H (1997) Biologische Invasionen: Das Beispiel *Mahonia aquifolium*. In: Feldmann R, Henle K, Auge H, Flachowsky J, Klotz S, Krönert R (eds) Regeneration und nachhaltige Landnutzung: Konzepte für belastete Regionen. Springer, Berlin, pp 124–129

Bakker PA, Boeve E (1985) Stinzenplanten. Uitgeverij Terra Zutphen

Bauer N (2005) Attitudes towards Wilderness and Public Demands on Wilderness Areas. In: Kowarik I, Körner S (eds) Urban Wild Woodlands. Springer, Berlin Heidelberg, pp 47–66

Baxter JW, Pickett STA, Dighton J, Carreiro MM (2002) Nitrogen and phosphorus availability in oak forest stands exposed to contrasting anthropogenic impacts. Soil Biology & Biochemistry 34(5):623–633

Bernatsky J (1904) Anordnung der Formationen nach ihrer Beeinflussung seitens der menschlichen Kultur und der Weidetiere. Engler's Bot Jb 94(1):1–8

Bork H-R, Bork H, Dalchow C, Piorr H-P, Schatz T, Schröder A (1998) Landschaftsentwicklung in Mitteleuropa. Klett-Perthes, Gotha

Box J (1993) Conservation or Greening? The Challenge of Post-industrial Landscapes. British Wildlife 4:273–279

Broggi MF (1992) Warum Naturwald? Gedanken zum naturgemässen Wald. Sihlwald Nachrichten 5: 4–41

Brothers TS, Spingarn A (1992) Forest fragmentation and alien plant invasion of central Indiana old-growth forests. Conservation Biology 6(1):91–100

Burckhardt D, Baur B, Studer A (eds) (2003) Fauna und Flora auf dem Eisenbahngelände im Norden Basels. Monographien der Entomologischen Gesellschaft Basel 1

Caneva G, Pacini A, Celesti Grapow L, Ceschin S (2003) The Colosseum's use and state of abandonment analysed through its flora. International Biodeterioration & Biodegradation 51:211–219

Celesti Grapow L, Blasi C (2003) Archaeological sites as areas for biodiversity conservation in cities: the spontaneous vascular flora of the Caracalla Baths in Rome. Webbia 58(1):77–102

Costanza R, dArge R, de Groot R, Farber S, Grasso M, Hannon B, Limburg K, Naeem S, ONeill RV, Paruelo J, Raskin RG, Sutton P, van den Belt M (1997) The value of the world's ecosystem services and natural capital Nature 387(6630):253–260

Davis BNK (1976) Wildlife, Urbanisation and Industry. Biol Conserv 10:249–291

De Jong E (1998) Der Garten als dritte Natur. In: Kowarik I, Schmidt E, Sigel B (eds) Naturschutz und Denkmalpflege. Wege zu einem Dialog im Garten. vdf Hochschulverlag, Zürich, pp 17–27

Dettmar J (1992) Industrietypische Flora und Vegetation im Ruhrgebiet. Diss Bot 191:1–397

Dettmar J (2005) Forests for Shrinking Cities? The Project "Industrial Forests of the Ruhr". In: Kowarik I, Körner S (eds) Urban Wild Woodlands. Springer, Berlin Heidelberg, pp 263–276

Dettmar J, Ganser K (1999) IndustrieNatur – Ökologie und Gartenkunst im Emscherpark. Ulmer, Stuttgart

Dickson JH (1989) Conservation and the Botany of Bings: Observations from the Glasgow Area. Trans Bot Soc Edinb 45:493–500

Durrant JO, Shumway JM (2004) Attitudes toward wilderness study areas: A survey of six southeastern Utah counties. Environmental Management 33(2):271–283

Ebrecht L, Schmidt W (2003) Nitrogen mineralization and vegetation along skidding tracks. Annals of Forest Science 60(7):733–740

Ellenberg H (1963) Vegetation Mitteleuropas mit den Alpen in kausaler, dynamischer und historischer Sicht. Ulmer, Stuttgart

Ellenberg H (1988) Vegetation ecology of Central Europe, 4th edn, Cambridge University Press, Cambridge

Ewert AW (1998) A comparison of urban-proximate and urban-distant wilderness users on selected variables. Environmental Management 22(6):927–935

Falinski JB (1966) Antropogeniczna Roslinnosc puszczy Bislowieskiej. PhD thesis, Univ Warszawa

Falinski JB (1986) Vegetation dynamics in temperate lowland primeval forests. Ecological studies in Bialowieza forest. Geobotany 8, Junk, Dordrecht Boston Lancaster

Fischer W (1975) Vegetationskundliche Aspekte der Ruderalisation von Waldstandorten im Berliner Gebiet. Arch Natursch u Landschaftsforsch 15(1):21–32

Freedman B, Love S, ONeil B (1996) Tree species composition, structure, and carbon storage in stands of urban forest of varying character in Halifax, Nova Scotia. Canadian Field-Naturalist 110(4):675–682

Géhu JM, Géhu J (1979) Essai d´évaluation phytocoenotique de l´artificialisation des paysages. Séminaire de Phytosociol appliquée, pp 95–118

Godefroid S, Koedam N (2003a) Identifying indicator plant species of habitat quality and invasibility as a guide for peri-urban forest management. Biodiversity and Conservation 12(8):1699–1713

Godefroid S, Koedam N (2003b) Distribution pattern of the flora in a peri-urban forest: an effect of the city-forest ecotone. Landscape and Urban Planning 65:169–185

Graae BJ (2002) The role of epizoochorous seed dispersal of forest plant species in a fragmented landscape. Seed Science Research 12:113–121

Harrison C, Davies G (2002) Conserving biodiversity that matters: practitioners' perspectives on brownfield development and urban nature conservation in London. Journal of Environmental Management 65(1):95–108

Harrison RP (1992) Forests. The shadow of civilization. University of Chicago Press, Chicago

Henderson N (1992) Wilderness and the nature conservation ideal. Britain, Canada, and the United States contrasted. Ambio 21(6):394–399

Hennebo D (1971) Sechshundert Jahre Eilenriede Hannover. Das Gartenamt 20(12):567–570

Hodkinson DJ, Thompson K (1997) Plant dispersal: the role of man. Journal of Applied Ecology 34(6):1484–1496

Hohmann U, Bartussek I (2001) Der Waschbär. Oertel and Spörer, Reutlingen

Jalas J (1955) Hemerobe und hemerochore Pflanzenarten. Ein terminologischer Reformversuch. Acta Soc Fauna Flora Fenn 72(11):1–15

Jim CY, Liu HT (2001) Species diversity of three major urban forest types in Guanzhou City, China. Forest Ecology and Management 146(1-3):99–114

Jochimsen M (1991) Ökologische Gesichtspunkte zur Vegetationsentwicklung auf Bergehalden. In: Wiggering H, Kerth M (eds) Bergehalden des Steinkohlenbergbaus. Vieweg, Wiesbaden, pp 155–162

Jorgensen A, Hitchmough J, Dunnett N (2005) Living in the Urban Wildwoods: A Case Study of Birchwood, Warrington New Town, UK. In: Kowarik I, Körner S (eds) Urban Wild Woodlands. Springer, Berlin Heidelberg, pp 95–116

Keil P, Sarazin A, Loos GH, Fuchs R (2002) Eine bemerkenswerte industriebegleitende Pteridophyten-Flora in Duisburg, im Randbereich des Naturraumes „Niederrheinisches Tiefland“. Decheniana 155:5–12

Kelcey JG (1975) Industrial Development and Wildlife Conservation. Environmental Conservation 2:99–108

Kendle T, Forbes S (1997) Urban nature conservation. Chapman and Hall, London

Kohler A, Sukopp H (1964) Über die Gehölzentwicklung auf Berliner Trümmerstandorten. Ber Dt Bot Ges 76(10):389–406

Körner S (2005) Nature Conservation, Forestry, Landscape Architecture and Historic Preservation: Perspectives for a Conceptual Alliance. In: Kowarik I, Körner S (eds) Urban Wild Woodlands. Springer, Berlin Heidelberg, pp 193–220

Kosmale S (1981) Die Wechselbeziehungen zwischen Gärten, Parkanlagen und der Flora der Umgebung im westlichen Erzgebirge. Hercynia NF 18(4):441–452

Kowarik I (1988) Zum menschlichen Einfluß auf Flora und Vegetation. Theoretische Konzepte und ein Quantifizierungsansatz am Beispiel von Berlin (West), Landschaftsentwicklung und Umweltforschung 56:1–280

Kowarik I (1991) Unkraut oder Urwald? Natur der vierten Art auf dem Gleisdreieck. In: Bundesgartenschau 1995 GmbH (ed) Dokumentation Gleisdreieck morgen. Sechs Ideen für einen Park, Berlin, pp 45–55

Kowarik I (1992a) Einführung und Ausbreitung nichteinheimischer Gehölzarten in Berlin und Brandenburg. Verh Bot Ver Berlin Brandenburg, Beiheft 3:1–188

Kowarik I (1992b) Das Besondere der städtischen Vegetation. Schriftenreihe des Deutschen Rates für Landespflege, 61:33–47

Kowarik I (1995) On the role of alien species in urban flora and vegetation. In: Pysek P, Prach K, Rejmanek M, Wade M (eds) Plant invasions. General aspects and special problems. SPB Academic Publ, Amsterdam, pp 85–103

Kowarik I (1998) Historische Gärten und Parkanlagen als Gegenstand eines Denkmal-orientierten Naturschutzes. In: Kowarik I, Schmidt E, Sigel B (eds) Naturschutz und Denkmalpflege. Wege zu einem Dialog im Garten. vdf Hochschulverlag, Zürich, pp 111–140

Kowarik I (1999) Natürlichkeit, Naturnähe und Hemerobie als Bewertungskriterien. In: Konold W, Böcker R, Hampicke U (eds) Handbuch Naturschutz und Landschaftspflege. V-2.1, Ecomed, Landsberg pp 1–18

Kowarik I (2003) Biologische Invasionen. Neophyten und Neozoen in Mitteleuropa. Ulmer, Stuttgart

Kowarik I (2005) Urban ornamentals escaped from cultivation. In: Gressel J (ed) Crop ferality and volunteerism. CRC Press, Boca Raton (in press)

Kowarik I, Langer A (1994) Vegetation einer Berliner Eisenbahnfläche (Schöneberger Südgelände) im vierten Jahrzehnt der Sukzession. Verh Bot Ver Berlin Brandenburg 127:5–43

Kowarik I, Langer A (2005) Natur-Park Südgelände: Linking Conservation and Recreation in an Abandoned Rail Yard in Berlin. In: Kowarik I, Körner S (eds) Urban Wild Woodlands. Springer, Berlin Heidelberg, pp 187–299

Kowarik I, Sukopp H (1984) Auswirkungen von Luftverunreinigungen auf die spontane Vegetation (Farn- und Blütenpflanzen). Angew Bot 58:157–170

Kowarik I, Körner S, Poggendorf L (2004) Südgelände: Vom Natur- zum Erlebnis-Park. Garten und Landschaft 114(2):24–27

Kowarik I Schmidt E, Sigel B (eds) (1998) Naturschutz und Denkmalpflege. Wege zu einem Dialog im Garten. vdf Hochschulverlag, Zürich

Kreh W (1955) Das Ergebnis der Vegetationsentwicklung auf dem Stuttgarter Trümmerschutt. Mitt Flor-Soz Arbeitsgem NF 5:69–75

Kunick W (1982) Zonierung des Stadtgebietes von Berlin (West). Ergebnisse floristischer Untersuchungen. Landschaftsentwicklung und Umweltforschung 14:1–164

Küster H (1998) Geschichte des Waldes. Von der Urzeit bis zur Gegenwart. Beck, München

Lehvävirta S, Rita H (2002) Natural regeneration of trees in urban woodlands. Journal of Vegetation Science 13 (1):57–66

Long G (1974) Diagnostic phyto-écologique et aménagement du territoire. Masson, Paris

Lutz AR, Simpson-Housley P, de Man AF (1999) Wilderness – Rural and urban attitudes and perceptions. Environment and Behavior 31:259–266

Malmivaara M, Lofstrom I, Vanha-Majamaa I (2002) Anthropogenic effects on understorey vegetation in Myrtillus type urban forests in southern Finland. Silva Fennica 36(1):367–381

Mantel K (1990) Wald und Forst in der Geschichte. Schaper, Alfeld Hannover

Marzluff JM (2001) Worldwide urbanization and its effects on birds. In: Marzluff JM, Bowman R, Donnelly R (eds) Avian ecology in an urbanizing world. Kluwer, Norwell, pp 19–47

McDonnell MJ, Pickett STA (1990) Ecosystem structure and function along urban-rural gradients: an unexploited opportunity for ecology. Ecology 71(4):1232–1237

McDonnell MJ, Rudnicky JL (1989) Forty-eight years of canopy change in a hardwood-hemlock forest in New York City. Bull Torrey Bot Club 116(1):52–64

McKinney ML (2002) Urbanization, biodiversity, and conservation. Bioscience 52(10):883–890

Miyawaki A, Fujiwara K (1975) Ein Versuch zur Kartierung des Natürlichkeitsgrades der Vegetation und Anwendungsmöglichkeiten dieser Karte für den Umwelt und Naturschutz am Beispiel der Stadt Fujisawa. Phytocoenologia 2(34):430–437

Moran MA (1984) Influence of adjacent land use on understory vegetation of New York forests. Urban Ecol 8:329–340

Mortberg UM (2001) Resident bird species in urban forest remnants; landscape and habitat perspectives Landscape Ecology 16(3):193–203

Müller M-L, Kaun A, Hard G (1978) Nutzerfrequenz und Wegrandvegetation. Zur Anwendung nonreaktiver Meßverfahren im Bereich anthropogener Vegetation. Landschaft + Stadt 10(4):172–179

Mulvaney M (2001) The effect of introduction pressure on the naturalization of ornamental woody plants in eastern Australia. In: Groves RH, Panetta FD, Virtue JG (eds) Weed Risk Assessment. CSIRO, Melbourne pp 186–193

Niemela J, Kotze DJ, Venn S, Penev L, Stoyanov I, Spence J, Hartley D, de Oca EM (2002) Carabid beetle assemblages (Coleoptera, Carabidae) across urban-rural gradients: an international comparison. Landscape Ecology 17(5):387–401

Ode AK, Fry GLA (2002) Visual aspects in urban woodland management. Urban Forestry & Urban Greening 1(3):15–24

Olaczek R (1982) Synanthropization of phytocoenoses. Memorabilia Zool 37:93–112

Parendes LA & Jones JA (2000) Role of light availability and dispersal in exotic plant invasion along roads and streams in the H. J. Andrews Experimental Forest, Oregon. Conservation Biology 14(1):64–75

Peterken GF (1974) A method of assessing woodland flora for conservation using indicator species. Biological Conservation 6:239–245

Peterken GF (1994) The definition, evaluation and management of ancient woods in Great Britain. NNA-Ber 7(3):102–114

Peterken GF, Game M (1984) Historical factors affecting the number and distribution of vascular plant species in the woodlands of central Lincolnshire. J Ecol 72:155–182

Platen R, Kowarik I (1995) Dynamik von Pflanzen-, Spinnen- und Laufkäfergemeinschaften bei der Sukzession von Trockenrasen zu Gehölzgesellschaften auf innerstädtischen Brachflächen in Berlin. Verh Ges f Ökol 24:431–439

Pott R, Hüppe J (1991) Die Hudelandschaften Nordwestdeutschlands. Abh Westfälisches Mus Naturkde 1/2:1–313

Prach K, Pysek P (1994) Spontaneous establishment of woody plants in central European derelict sites and their potential for reclamation. Restoration Ecology 2(3):190–197

Preston CD, Pearman DA, Dines TD. (eds.) 2002. New Atlas of the British and Irish Flora. Oxford University Press, Oxford

Rebele F (2003) Sukzessionen auf Abgrabungen und Aufschüttungen – Triebkräfte und Mechanismen. Ber Inst Landschafts- Pflanzenökologie Univ Hohenheim, Beiheft 17:67–92

Rebele F, Dettmar J (1996) Industriebrachen. Ökologie und Management. Ulmer, Stuttgart

Reidl K, Dettmar J (1993) Flora und Vegetation der Städte des Ruhrgebietes, insbesondere der Stadt Essen und der Industrieflächen. Ber z Dt Landeskde 67(2):299–326

Rejmanek M (2000) Invasive plants: approaches and predictions. Austral Ecol 25:497–506

Ringenberg J (1994) Analyse urbaner Gehölzbestände am Beispiel der Hamburger Wohnbebauung. Dr. Kovac, Hamburg

Rink D (2005) Surrogate Nature or Wilderness? Social Perceptions and Notions of Nature in an Urban Context. In: Kowarik I, Körner S (eds) Urban Wild Woodlands. Springer, Berlin Heidelberg, pp 67–80

Roovers P, Hermy M, Gulinck H (2002)Visitor profile, perceptions and expectations in forests from a gradient of increasing urbanisation in central Belgium. Landscape and Urban Planning 59(3):129–145

Rösler M (2004) Mit dem Kinderwagen in den Urwald. Ein Wildnisprojekt im Ballungsraum Saarbrücken geht neue Wege. Garten + Landschaft Heft 2:33–35

Sachse U (1989) Die anthropogene Ausbreitung von Berg- und Spitzahorn. Ökologische Voraussetzungen am Beispiel Berlins. Landschaftsentwicklung und Umweltforschung 63:1–132

Sachse U, Starfinger U, Kowarik I (1990) Synanthropic woody species in the urban area of Berlin (West). In: Sukopp H, Hejny S, Kowarik I (eds) Urban ecology. SPB Academic Publishing, The Hague, pp 233–243

Schlüter H (1984) Kennzeichnung und Bewertung des Natürlichkeitsgrades der Vegetation. Acta Bot Slov Acad Sci Sclovacae Ser A 1:277–283

Schlüter H (1992) Vegetationsökologische Analyse der Flächennutzungsmosaike Nordostdeutschlands. Natürlichkeitsgrad der Vegetation in den neuen Bundesländern. Naturschutz u Landschaftsplanung 5:173–180

Seibert P (1980) Ökologische Bewertung von homogenen Landschaftsteilen, Ökosystemen und Pflanzengesellschaften. Ber ANL 4(10):10–23

Seidling W (1999) Spatial structures of a subspontaneous population of *Taxus baccata* saplings. Flora 194:439–451

Shaw DC, Bible K (1996) Overview of forest canopy ecosystem functions with reference to urban and riparian systems. Northwest Science 70:1–6

Spethmann, W (1995) In-situ / ex-situ-Erhaltung von heimischen Straucharten. In: Kleinschmit J, Begemann F, Hammer K (eds) Erhaltung pflanzengenetischer Ressourcen in der Land- und Forstwirtschaft. Schriften zu Genetischen Ressourcen (ZADI, Bonn) 1:68–87

Sukopp H (1969) Der Einfluß des Menschen auf die Vegetation. Vegetatio 17:360–371

Sukopp H (1972) Wandel von Flora und Vegetation in Mitteleuropa unter dem Einfluß des Menschen. Ber Landw 50:112–139

Sukopp H, Scholz H (1964) *Parietaria pensylvanica* Mühlenb. ex Willd. in Berlin. Ber Dt Bot Ges 77:419–426

Tara K, Zimmermann K (1997) Brachen im Ruhrgebiet. LÖBF-Mitt 3/1997:16–21

Tarvainen O, Markkola AM, Strommer R (2003) Diversity of macrofungi and plants in Scots pine forests along an urban pollution gradient. Basic and Applied Ecology 4(6):547–556

Tischew S, Lorenz A (2005) Spontaneous Development of Peri-Urban Woodlands in Lignite Mining Areas of Eastern Germany. In: Kowarik I, Körner S (eds) Urban Wild Woodlands. Springer, Berlin Heidelberg, pp 163–180

Tyrvainen L (2001) Economic valuation of urban forest benefits in Finland. Journal of Environmental Management 62(1):75–92

Tyrvainen L, Väänänen H (1998) The economic value of urban forest amenities: an application of the contingent valuation method, Landscape and Urban Planning, 43:105–118

Venn SJ, Kotze DJ, Niemela J (2003) Urbanization effects on carabid diversity in boreal forests. European Journal of Entomology 100(1):73–80

Verheyen K, Bossuyt B, Hermy M, Tack G (1999) The land use history (1278-1990) of a mixed hardwood forest in western Belgium and its relationship with chemical soil characteristics Journal of Biogeography 26(5):1115–1128

Verheyen K, Bossuyt B, Honnay O, Hermy M (2003) Herbaceous plant community structure of ancient and recent forests in two contrasting forest types. Basic and Applied Ecology 4(6):537–546

von Hornstein F (1950) Theorie und Anwendung der Waldgeschichte. Forstwiss Centr bl 69:161–177

von Hornstein F(1954) Vom Sinn der Waldgeschichte. Angew Pflanzensoz 2:685–707
Walther G-R (1999) Distribution and limits of evergreen broad-leaved (laurophyllous) species in Switzerland. Botanica Helvetica 109:153–167
Webb SL, Kaunzinger CK (1993) Biological invasion of the Drew-University (New Jersey) forest preserve by Norway maple (*Acer platanoides* L.). Bull. Torrey Bot. Club 120:343–49
Weiss J (2003) „Industriewald Ruhrgebiet". Freiraumentwicklung durch Brachensukzession. LÖBF-Mitt 1/2003:55–59
Weiss J, Burghardt W, Gausmann P, Haag R, Haeupler H, Hamann M, Leder B, Schulte A, Stempelmann I (2005) Nature Returns to Abandoned Industrial Land: Monitoring Succession in Urban-Industrial Woodlands in the German Ruhr. In: Kowarik I, Körner S (eds) Urban Wild Woodlands. Springer, Berlin Heidelberg, pp 143–162
Wendland F (1979) Berlins Gärten und Parke. Von der Gründung der Stadt bis zum ausgehenden neunzehnten Jahrhundert. Fröhlich und Kaufmann, Berlin
Westhoff V (1949) Schaakspel met de natuur. Natuur Landschap 3
Westhoff V (1951) De beteknis van natuurgebieden voor wetenschap en praktijk. Contact-Comm Natuur- en Landschapsbescherming
Wulf M (2003) Preference of plant species for woodlands with differing habitat continuities. Flora 198:444–460 – im Text 2004
Wulf M, Kelm H-J (1994) Zur Bedeutung „historisch alter Wälder" für den Naturschutz - Untersuchungen naturnaher Wälder im Elbe-Weser-Dreieck. NNA-Ber 7(3):15–49
Zerbe S (2004) Influence of historical land use on present-day forest patterns: A case study in south-western Germany. Scandinavian Journal of Forest Research 19(3):261–273
Zisenis M (1996) Secondary woodland on Nunhead Cemetery, London, UK. Landschaftsentwicklung und Umweltforschung 104:73–79

New Perspectives for Urban Forests: Introducing Wild Woodlands

Cecil C. Konijnendijk

Danish Centre for Forest, Landscape and Planning

Nature, forests, trees and cities

Although antagonists by definition, nature and cities have had a much more complex relationship. Urbanisation has meant that natural areas have become cultivated and often overexploited and that nature has been removed as a dominant factor in the daily life of an increasing number of people. But this also triggered a longing to get back to nature and a desire to bring nature back to cities. Nature, though often in a cultivated form, was seen as having its place in cities and towns, for example, for aesthetic reasons. This notion was supported by ancient Greek and Roman civilisations, and later also during the Renaissance. During the era of Romanticism, more "pure" and untouched forms of nature were also sought (e.g. Lawrence 1993).

Increased threats to natural areas also led to the emergence of the nature-conservation movement. In Europe, this movement had its roots in the cities of the second half of the 19th century, where artists, scholars and the bourgeoisie undertook actions to preserve the remnants of once abundant natural areas. Often these early nature conservationists directed their attention towards forests close to cities and towns. Well-known examples are those of Fontainebleau, south of Paris, the Zonienwoud at the borders of Brussels, and the Sihlwald near Zurich (Konijnendijk 1999). In the case of Fontainebleau, which had gradually become a "Parisian promenade" after a railroad connection with Paris had been established, a group of painters and artists raised concerns about the many threats to the forest. This group, known as the School of Barbizon, raised awareness about the high natural and cultural-historical value of Fontainebleau and helped extend its protected status (INRA 1979).

But efforts to protect remnants of forests in and near cities date back even further. European "city-forest" history has many examples of city

Kowarik I, Körner S (eds) Wild Urban Woodlands.

administrations undertaking efforts to preserve nearby forests, primarily for the benefits of their residents. Examples are those of the Eilenriede Forest near Hanover, the Sihlwald close to Zurich, and Epping Forest of London (see Konijnendijk 1999 for these and other examples).

Forests and other tree resources have played a key role throughout the development of European cities and towns. Over time they have had different functions, starting with the provision of food, fuelwood, timber and fodder during medieval times. They provided the nobility with much-appreciated nearby hunting areas. Later their aesthetic and recreational values took priority, as forests and parks offered "healthy" options for spending increasing leisure time, and trees helped beautify grim urban living conditions. The environmental functions of trees in terms of providing shade, moderating the urban mesoclimate, helping to reduce air pollution, and protecting water resources have partly been known for some time, but are very much in focus today (Miller 1997).

With the large majority of Europeans now living in urban areas, policy-makers are stressing the importance of sustainable urban development, and of a high quality of urban life and urban environment (e.g. European Commission 1999). More structured attention to urban nature and to natural processes as integral parts of cities dates back more than twenty years (Goode 1995). The development of urban ecology (e.g. Duvigneaud 1974), for example, was one of the first of several integrative approaches aimed at the development and management of urban green structures.

Urban forestry was introduced as one of these integrative approaches, having gradually obtained a broader scientific and practical following. Urban forests as city-wide structures of all tree stands and individual trees were then introduced as important tools for maintaining or bringing back nature and "the wild" to urban residents' doorsteps. But what does "wild" mean within an urban-forestry context? The chapter discusses how careful consideration needs to be given to the relationship between people and trees when defining and working with "the wild urban woodland" as a concept.

Urban forestry as an integrative framework

Urban forestry has been defined as the art, science and technology of managing trees and forest resources in and around urban community ecosystems for the physiological, sociological, economic, and aesthetic benefits trees provide society (Helms 1998). The term was first coined in Canada as part of the title of a 1965 graduate study on municipal tree planting. In spite of initial resistance to the term from foresters (who doubted forestry's role in urban areas) as well as from other professions traditionally dealing with urban green space, it gradually found a broad following in North America. The emergence of urban forestry has been regarded as a reaction to eminent threats to urban trees—for example, those caused by introduced pests and diseases—which called for more integrative tree management approaches (Johnston 1996; Miller 1997).

Although the term "urban forestry" was introduced to the United Kingdom during the early 1980s, broader acceptance in Europe was not gained until the mid-1990s. While the *concept* of integrative planning and management of forest and other tree resources soon found support among European experts, implementation of the *term* proved more problematic, as the equivalent to "urban forest" in many European languages—for example, *Stadtwald* in German and *stadsbos* in Dutch—has traditionally referred to forest stands owned by cities and towns (Konijnendijk 2003).

Urban forestry has a broader scope, however, dealing with forest stands as well as groups of trees and individual trees. The British National Urban Forestry Unit has defined "the urban forest" as collectively describing all "trees and woods in an urban area: in parks, private gardens, streets, around factories, offices, hospitals and schools, on wasteland and in existing woodlands" (NUFU 1999). Forest stands within an urban-forestry context have come to be referred to as "urban woodlands" in order to make a distinction with the overall urban-forest concept (e.g. Bell et al. in press; Konijnendijk 2003).

During the first four decades of its history, urban forestry has also been given more normative content, focusing on aspects of multifunctionality and community participation (e.g. Johnston 1996) in an attempt to demonstrate its particular strengths and also its value for practical implementation. A study of the literature has resulted in the following synopsis of the main characteristics and strengths of the urban-forestry concept (based on Konijnendijk 2003):

- Urban focus, recognising and valuing rather than combating the challenges posed by urban societies and urban environments.
- Integrative approach, incorporating different elements of urban green structures into a whole (the "urban forest"), and ranging from technical to strategic dimensions of natural resource management.
- Strategic perspective, aimed at developing longer-term policies and plans for urban-tree resources, connecting to different sectors, agendas and programmes.
- Inter-/multidisciplinary character, involving experts from applied, natural as well as social sciences.

- Multifunctional focus, stressing the socio-cultural, environmental as well as economic benefits and services urban forests can provide.
- Participatory approach, aimed at developing partnerships between all community stakeholders.

The latter two characteristics have been central to the development of urban forestry. The provision of multiple benefits—not least of all, social ones—to local communities is its raison d'être. In terms of urban forestry's multifunctional focus, the multiple social services provided by urban forests are given priority. These include, among others, providing outdoor recreational environments and pleasant living and working environments to urban residents, thus contributing to human health and well-being (e.g. Kuo 2003). As mentioned above, various environmental or ecological services related to urban mesoclimate, air-pollution reduction and watershed management have also come into focus. The ecological dimension relates to conserving biodiversity values and representing nature in human-made environments as well.

Through its multifunctional and in particular its participatory character, urban forestry relates strongly to the, perhaps better known, concept of community forestry. Most commonly applied in the rural context of developing countries, community forestry has been defined as any form of social forestry based on the local people's direct participation in the production process, either by growing trees themselves or by processing tree products locally (Raintree 1991). But community forestry is increasingly also applied to more urban settings, as in the case of the English Community Forest programme for environmental and socio-economic regeneration of twelve large urban agglomerations (e.g. Johnston 1996). Brender and Carey (1998) provide a broader perspective of community forestry, stressing its focus on sustainable forestry for community well-being, and with key attributes that include residents' access to land and its resources, as well as local involvement in decision-making related to the forest.

Various authors have emphasised the importance of placing local residents (i.e. local forest users) centrally in urban green-space planning and management. Rachel and Stephen Kaplan (1989), for example, mention the importance of close links between people and nearby nature. These links may be "special spots" a person (or group of people) is very possessive about and considers his or her own (Frey 1981; cited in Kaplan and Kaplan 1989). Environmental psychology provides a range of theories, such as that of "place identity" (Proshansky et al. 1983), that give insight into close relationships between people and green space and explain the intensity of social conflicts that frequently emerge, e.g. over the cutting of urban trees. When attempting to define the "woodland component" of the urban forest, Konijnendijk (1999) stressed the dominance of local urban actors and their interests, values and norms in their use and related decision-making processes.

"Wild urban woodlands": placing people's perceptions and preferences first

Expert debate on "wild woodlands"

Multifunctionality in urban forestry is about opting for the right combination of urban-forest functions in the right place. Limited urban-forest resources have to meet the high and diverse demands of thousands and sometimes millions of local users. Combining social and ecological demands is a key task from the perspective of sustainable urban-forest management (e.g. Volk 1995). In this context, the meaning and role of "wild urban woodland" as a concept should be discussed (see also Kowarik 2005).

According to Merriam-Webster's online dictionary of English (Merriam-Webster 2003), "wild" as an adjective can mean "living in a state of nature and not ordinarily tame or domesticated" or "growing or produced without human care or aid". With reference to land, "wild" relates to "not inhabited or cultivated" or "not amenable to human habitation or cultivation" (as in "wilderness").

Using the term "wild" to refer to the absence of human interference or cultivation in an urban, highly cultivated context may seem problematic. On the other hand, experience has shown that there is room for natural processes as well as original natural vegetation in urban areas (Goode 1995), for example, in the case of vegetation colonisation on former industrial land (Kowarik 2005).

Expert debate on the meaning of "natural" in a general forestry context is ongoing. Peterken (1995), for example, refers to a spectrum between totally natural and totally artificial when classifying woodlands. The term "semi-natural woodland" is often used in a British context, referring, for example, to woodlands that resemble those that would emerge after natural colonisation or succession (while recognising a certain amount of human influence). Also applied is the term "ancient woodland", which according to Rackham (1990), refers to woodlands dating back at least to 1600. In the German-language literature, the primeval forest is mostly described as *Urwald* (Senatsverwaltung… 2001).

Bell et al. (in press) discuss difficulties in distinguishing between "forest", "woodland" and "park" in an urban-forestry context, with "forest" traditionally referring to unenclosed, wild areas; "park" to unclosed, managed areas; and "woodlands" to something between the two. Another related debate involves indigenous and exotic vegetation.

Kowarik (2005) differentiates among four types of "nature" when studying, for example, urban woodlands. These types differ in terms of the level of human interference and similarity to original vegetation. "Nature of the first kind" includes all natural, untouched woodlands, while natural processes are also central to "nature of the fourth kind" which relates to woodlands that have developed spontaneously on urban-industrial sites.

These discussions, however, are very much expert-based, focusing on ecological and management considerations, and closely linking "wild" with "natural". As we have seen, however, a key characteristic of urban forestry is that it places its primary beneficiaries, i.e. local resident communities, centrally. Benefits are not only generated *for* local communities, but the aim is often also to achieve *empowerment* of these communities in urban-forestry decision-making. If these local residents play such a crucial role, what then does the "wild urban woodland" mean to them?

"Wild urban woodland": user perceptions and preferences

Studies from around the world (e.g. Kaplan and Kaplan 1989; Van den Berg and De Vries 2000) show that residents prefer urban environments with nature over those without nature. Experiencing and enjoying nature is often a main motivation for visiting forests and other green areas, urban as well as rural. A nation-wide survey of the Dutch population (Nas 1998), for example, showed that 91% of respondents mentioned "enjoying natural beauty" as a main argument for visiting the Dutch forests, many of which can be considered "urban". In terms of perceived natural value, larger forest and nature areas outside cities tend to "score" higher in perception studies than inner-city green areas (Van den Berg and De Vries 2000). Wiggers and Gadet (1996) showed that the residents of Amsterdam eagerly use green spaces when these are present in their neighbourhood, but also that inner-city green has primarily become a setting for social activities among young singles rather than being appreciated for its natural values. Thus preferences with regards to urban-forest benefits differ according to social and cultural differences, as well as location.

That urban residents hold very different interpretations of the meaning of "wild" and "nature/natural" in an urban-forestry context is illustrated by two examples from outside Europe. A resident focus-group study in the city of Christchurch, New Zealand (Kilvington and Wilkinson 1999) on community attitudes toward natural vegetation in the urban environment showed that the terms "wild" and "natural" had highly confused meanings for many participants. A focus-group study among youths in the city-state of Singapore (Kong et al. 1999) found this diversity of attitude and perception as well, as nature was described both as "unpredictable and dangerous" (with reference to faraway places) and as "safe and fun" (when referring to the orderliness of well-maintained nature in the city). Most young Singaporeans interviewed showed little interest in or affinity with nature, most likely because of limited daily contact with nature as a "familiar" part of the living environment.

When users' preferences of urban forests are considered, factors other than "naturalness" (in terms of absent or limited human interference) may be more important. Roovers et al. (2002) studied visitor perceptions and expectations towards forests in central Belgium, moving along a gradient of increasing urbanisation. A majority of the visitors (59%) surveyed preferred mixed forests, and 78.9% strongly favoured some or high variation in forest layers. Diversity was also appreciated in terms of variation in topography. But these aspects are not unique to "natural" or "wild" forests from an expert or ecological perspective. Interviewed residents of the Bulgarian city of Stara Zagora, for example, described exotic cedar stands in a local woodland park as being the closest to their perception of the Bulgarian "natural forest" (Van Herzele, personal communication).

Recognising local users' perceptions and preferences of "wild urban woodlands" is necessary as part of the urban-forestry approach, but will not make the debate less complicated. "Wild" can have many different meanings to people, as has been briefly illustrated here. In line with different perceptions of "wild urban woodlands", different roles and benefits will be in focus.

Benefits and drawbacks of wild urban woodlands

Benefits of wild urban woodlands

In the light of the above considerations, biodiversity and "naturalness" aspects may seem less central to urban forestry, where social and environmental services are favoured. Studies have shown, however, that urban green space can support significant biodiversity, for example, in terms of habitat and species diversity (e.g. Sukopp and Werner 1987). In recognition of biodiversity and high natural values, sometimes in combination with cultural-historical importance, many major European cities, such as

Moscow, Stockholm, Vienna and Warsaw, host national parks and nature reserves within their boundaries.

The primary role of urban forests, however, is a social one, i.e. to provide attractive environments for urban dwellers to live, work, and spend their leisure time. Among all forests, those situated closest to cities are by far the most visited, as evidence from Sweden shows (Rydberg 2001). Visitor numbers for urban woodlands may be in the thousands per hectare on an annual basis (see Konijnendijk 1999 for examples). Urban forests provide people with the opportunity to get away from hectic urban life and to experience nature. Urban forests are also areas of fun and play. Braun (1998) relates changing lifestyles to changing outdoor recreational patterns, e.g. demonstrating that more adventurous types of forest recreation, such as mountain biking, have become more popular. The need for children to have areas to play and experience nature has led to the establishment of special "play forests", for example, in Sweden (Rydberg 2001). "Wild" thus does not necessarily have to relate to natural *per se*, but rather to a forest that users' perceive as being wild and thus suitable for their preferred use in terms of variation, topography, lack of human artefacts and visual interference, and so forth.

However, the "wild" element of urban forests in terms of their natural values and natural processes – with or without human interference – can certainly also be of social importance. Trees, for example, indicate the changing of seasons. The Dutch "Nature Calendar" initiative (www.natuurkalender.nl; in Dutch only) has linked up to this daily relationship between people and nature in their immediate living environment. Residents are asked to enter their first sightings of selected bird species, flowering of plants, and so forth on a special website. The researchers behind the site use the information for phenological studies, e.g. of the impacts of climate change on Dutch nature. But enhancing people's everyday involvement with nature is a clear secondary objective. Having "wild woodlands" where there is room for natural processes is also important in cities from a nature education perspective (e.g. Rydberg 2001). Many of Europe's urban woodlands host "forest schools" where school classes can spend a day or more learning about nature (e.g. Konijnendijk 1999).

Disadvantages of "wild urban woodlands"

Having the "wild" represented in cities through forests and trees can also cause problems which need to be taken into account by forest managers. Problems with trees and woodlands in and near cities range from wildfire hazards and allergy problems to nuisances caused by falling leaves and fruits.

Studies have indicated that residents' sense of safety can be negatively affected by reduced visibility caused by abundant vegetation and undergrowth in urban green areas (e.g. Nibbering and Van Geel 1993; Burgess 1995), although Kuo (2003) provides evidence from the United States that people living in greener surroundings felt safer and better adjusted than those living next to barren areas. Other drawbacks are associated with wild animals being present in cities, even in the heart of a metropolis such as New York (Blumenthal 2003). The animals themselves can cause damages, for example, to private gardens, as in the case of wild boar roaming the streets of Berlin, or by carrying diseases such as Lyme's disease.

Perspective: developing wild urban woodlands

In spite of the drawbacks outlined above, it is generally accepted that nature also has its place in cities, not least of all represented by trees and woodlands. The question of what form of nature is represented—ranging from highly cultivated and controlled to "wild" meaning with limited or no human interference—is a more difficult one. The fact that urban residents' complex perceptions of what is "wild" must be taken into account does not make the work of green-space planners and managers easier. One the other hand, the multiple social as well as ecological meanings of "wild" create opportunities for urban forestry.

In urban areas where demands and pressures are high and space for nature is limited, some form of management seems necessary for most parts of the urban forest. In order to accommodate social as well as ecological considerations, the concept of close-to-nature forest management has gradually become popular in urban forestry (e.g. Senatsverwaltung... 1991; Jacsman 1998; Otto 1998). Developed as early as during the 1880s by Gayer, close-to-nature forestry (*naturgemäße Waldbewirtschaftung / Waldwirtschaft)* aimed at combining nature conservation and landscape objectives with those of forestry and timber production (Senatsverwaltung... 1991). Close-to-nature forestry aims to imitate nature in management and provide more room for natural processes.

Still, in order to serve multifunctional needs, choices will need to be made, as not all functions can be realised in every part of the urban forest. Through zoning, urban foresters across Europe have attempted to conserve areas – for example, in the central part of more extensive urban woodland areas – where natural values are given priority and nature is given more freedom. This does not necessarily clash with the need to place the values of local urban communities centrally. These "nature cores" or "wild woodlands" also provide significant social benefits, for example by offering a setting for nature-oriented recreation, contemplation, and nature education. Appropriate, if only minimal, recreational facilities can help alleviate visitor pressure (Jacsman 1998). Starting from highly cultivated elements of the urban forest such as gardens and street trees, a "story" of increasing naturalness and "wildness" and the need to look at urban green structures as an integrated whole can be told from people's doorsteps all the way to the heart of the urban woodlands and nature areas.

In their efforts to introduce a greater focus on natural values and processes, forest managers benefit from the fact that visitors have shown themselves more appreciative of closer-to-nature management. A comparison of visitor preference studies carried out in Sweden in 1977 and 1997 showed that photographs of natural forests with fallen-down trees and dead wood were rated significantly higher in 1997 (Lindhagen and Hörnsten 2000). A survey among residents of municipalities in the northern part of the Netherlands showed the residents to be rather positive about the introduction of "ecological management" of municipal green space, for example, in terms of grazing, less mowing and refraining from using pesticides and fertilisers (Wolterbeek 1999). Restrictive measures enhancing natural value are often accepted by visitors, as in the case of visitors to a peri-urban woodland and recreation area near Hanover, Germany. Ninety percent of the visitors were found to accept limitations and restrictions serving nature-conservation objectives (Mehls and Krischer 1996).

Proper communication and information about the efforts to enhance natural values seem crucial for further enhancing public understanding and awareness. Photos of a "natural forest" shown to Danish forest visitors were given a higher ranking as soon as the caption "natural forest" was added, indicating the importance of providing information (Jensen 2000). The selection of appropriate, non-intrusive methods of providing information proved to be important, as panels and signs were more popular among the Danish interviewees than exhibitions and forest rangers. When a serious conflict between residents and foresters emerged over forest management practices in the Mastbos, an urban woodland near Breda, the Netherlands, the role of an external expert who introduced residents to various management practices and even involved them in management proved crucial for conflict management (Konijnendijk 1995). More active involvement of urban residents is also envisioned in Singapore in order to re-establish a relationship between urban youth and nature. Activities such as educational visits and cultural performances have been suggested, as have efforts to turn green areas into places of everyday use (Kong et al. 1999).

In conclusion, there is definitely a role for the "wild woodland" in an urban-forestry context, as long as different meanings of "wild" based on residents' perceptions and preferences are considered, and management measures favouring nature and natural processes are properly communicated.

References

Bell S, Blom D, Rautamäki M, Castel Branco C, Simson A, Olsen, IA (in press) The design of the urban forest. In: Nilsson K, Randrup TB, Konijnendijk CC (eds) Urban forests and trees in Europe. Springer Academic, Heidelberg

Blumenthal R (2003) Gun-toting sea lions and other park tales. New York Times, Sunday, 1 June 2003:5

Braun B (1998) Walderholung im Spiegel der Sozialwissenschaften. Forstwissenschaftliches Centralblatt 117:44–62

Brender T, Carey H (1998) Community forestry defined. Journal of Forestry 96(3):21–23

Burgess J (1995) 'Growing in confidence' – Understanding people's perceptions of urban fringe woodlands. Technical Report. Countryside Commission, Cheltenham

Duvigneaud P (1974) La synthèse écologique: populations, communautés, écosystèmes, biosphère, noosphère. Doin, Paris

European Commission (1999) Sustainable urban development in the European Union: A framework for Action. COM(98) 605. Commission of the European Communities, Brussels

Goode DA (1995) Het ontwerpen van natuur in stadsparken (Designing nature in urban parks) (in Dutch). Groenkontakt 21(4):44–48

Helms JA (1998) The dictionary of forestry. Society of American Foresters, Bethesda

INRA (1979) La forêt et la ville: essai sur la forêt dans l'environnement urban et industriel. Institut Nationalde Recherche Agronomique, Station de recherches sur la fôret et l'environnement, Versailles

Jacsman J (1998) Konsequenzen der intensiven Erholungsnutzung für die Wälder im städtischen Raum. Schweizerische Zeitschrift für Forstwesen 149(6):423–439

Jensen FS (2000) The effects of information on forest visitors' acceptance of various management actions. Forestry 73(2):175–172

Johnston M (1996) A brief history of urban forestry in the United States. Arboricultural Journal 20:257–278

Kaplan R, Kaplan S (1989) The experience of nature. Cambridge Univ. Press, Cambridge

Kilvington M, Wilkinson R (1999) Community attitudes to vegetation in the urban environment: a Christchurch case study. Landscape Research Science Series No. 22. Manaaki Whenua Press, Lincoln

Kong L, Yuen B, Sodhi N, Briffet C (1999) The construction and experience of nature: perspectives from urban youths. Tijdschrift voor Economie en Sociale Geografie 90(1):3–16
Konijnendijk CC (1995) Het Mastbos heeft zijn vrienden gevonden: conflict over te voeren beheer lijkt ten einde (The Mastbos has found its friends: conflict over envisaged management seems to have ended) (in Dutch). Nederlands Bosbouwtijdschrift 67(2):61–67
Konijnendijk CC (1999) Urban forestry in Europe: A comparative study of concepts, policies and planning for forest conservation, management and development in and around major European Cities. Ph.D. Thesis, Research Notes No. 90. Faculty of Forestry, University of Joensuu
Konijnendijk CC (2003) A decade of urban forestry in Europe. Forest Policy and Economics 5(3):173–186
Kowarik I (2005) Wild urban woodlands: Towards a conceptual framework. In: Kowarik I, Körner S (eds) Urban Wild Woodlands. Springer, Berlin Heidelberg, pp 1–32
Kuo FE (2003) The role of arboriculture in a healthy social ecology. Journal of Arboriculture 29(3):148–155
Lawrence HW (1993) The Neoclassical origins of modern urban forests. Forest and Conservation History 37(1):26–36
Lindhagen A, Hörnsten L (2000) Forest recreation in 1977 and 1997 in Sweden: changes in public preferences and behaviour. Forestry 73(2):143–153
Mehls H-H, Krischer V (1996) Die Erholungslandschaft 'Grosser Deister' – Eine Befragung der Waldbesucher zur Erholungsleistung. Forst und Holz 51(5):142–144
Merriam-Webster (2003). Merriam-Webster Online Dictionary, www.m-w.com. Accessed on 22 December 2003
Miller RW (1997) Urban forestry: planning and managing urban green spaces, 2nd edn. Prentice Hall, Upper Saddle River, NJ
Nas R (1998) Gedrag, beleving en wensen van de Nederlander ten aanzien van het bos (Behaviour, perception and wishes of the Dutchman regarding forest) (in Dutch). Nederlands Bosbouwtijdschrift 70: 247-250
Nibbering C, Van Geel (1993) Zucht naar de kathedraal: beleving en beheer in de Haarlemmerhout (Longing for the cathedral: perception and management in the Haarlemmerhout) (in Dutch). Groen 49(12):26–31
NUFU (1999) Trees and woods in towns and cities. How to develop local strategies for urban forestry. National Urban Forestry Unit, Wolverhampton
Otto HJ (1998) Stadtnähe Wälder – kann die Forstwirtschaft sich noch verständlich machen? Forst und Holz 53(1):19–22
Peterken G (1995) Natural woodland: Ecology and conservation in northern temperate regions. Cambridge University Press, Cambridge
Proshansky HH, Fabian AK, Kaminoff R (1983) Place-identity: Physical world socialization of the self. Journal of Environmental Psychology 3(1):57–83
Rackham O (1990) Trees and woodland in the British landscape: the complete history of Britain's trees, woods and hedgerows, revised edition. Phoenix Press, London

Raintree JB (1991) Socioeconomic attributes of trees and tree planting practices. FAO (with ICRAF), Rome

Roovers P, Hermy M, Gulinck H (2002) Visitor profile, perceptions and expectations in forests from a gradient of increasing urbanisation in central Belgium. Landscape and Urban Planning 59:129–145

Rydberg D (2001) Skogens sociale varden (The forest's social value) (in Swedish). Rapport 8j 2001. Skogsstyrelsen, Jönköping

Senatsverwaltung für Stadtentwicklung und Umweltschutz Berlin (1991) Vom Kulturwald zum Naturwald: Landschaftspflegekonzept Grunewald. Arbeitsmaterialen der Berliner Forsten 1, Berlin

Sukopp H, Werner P (1987) Development of flora and fauna in urban areas. Council of Europe, Strasbourg

Van den Berg AE, De Vries S (2000) Het binnenstedelijke buitengevoel (The innercity outdoor feeling) (in Dutch). Levende Natuur 101:182–185

Volk H (1995) Wald und Erholung im Wertewandel. Beiträge zur Forstwirtschaft und Landschaftsökologie 29(3):126–133

Wiggers R, Gadet J (1996) Het grote groenonderzoek: Het bezoek aan en gebruik van parken, recreatiegebieden en groen in de woonomgeving in Amsterdam (The big green study: Visiting and use of parks, recreation areas and green in the living environment in Amsterdam) (in Dutch). Stedelijk Beheer, Amsterdam

Wolterbeek T (1999) Bewoners vaak positief over ecologisch beheer (Residents often positive about ecological management) (in Dutch). Tuin & Landschap 21(1999):58–59

Attitudes towards Wilderness and Public Demands on Wilderness Areas

Nicole Bauer

Swiss Federal Research Institute WSL, Section Landscape and Society

Introduction

In recent years the general conditions for land use in Switzerland for agriculture and forestry have changed as less area is needed for farming. Primary production is no longer profitable and state subsidies have been reduced. This means that decisions will have to be made about whether to abandon the cultivation of many areas of land now used for agriculture and forestry. In addition, some nature conservation organisations are lobbying to stop many areas of land from being used for agriculture and forestry and to have them designated as conservation areas.

As a result of these developments, wilderness and its spread have been the subject of debate in Switzerland for some years now. The main focus of discussion has not, however, been primary wilderness, i.e. the state that prevails when there has never been any visible human intervention. Rather the focus has been much more on the spread of secondary wilderness, which develops when the controlling human interventions are discontinued and the land returns to wilderness. The debate has involved political decision makers, nature conservation organisations, and local stakeholders. To ensure that decisions about the future development of the landscape are also carried by the general public, it is important to discover the basis of the public's support for, or its opposition to, wilderness and its spread. What is involved besides or in addition to expert knowledge, and what conceptions of wilderness and wilderness areas do people have?

What do we know?

There have been an increasing number of papers and conferences on the topic of wilderness in recent years, which indicates that debates about wil-

Kowarik I, Körner S (eds) Wild Urban Woodlands.
 pp 47–66

derness are not confined to Switzerland. Researchers and representatives with different disciplinary backgrounds from state and local nature conservation offices have met to discuss the terms "wilderness" (*Wildnis*) and "wilderness spread" (*Verwilderung*: Trommer 1997; Broggi 1999), and the opportunities which derelict land areas present as space for nature to develop free from human influence.

Research on the relationship between humans and nature, on perceptions of nature and how people experience it has a long tradition in the social sciences. In this paper it will only be possible to elucidate a few studies we believe to be particularly relevant to this investigation.

Kellert (1980) distinguishes, on the basis of empirical research, nine types of attitudes people have towards nature:

- the utilitarian type, which views nature in terms of its material value
- the naturalistic type, which regards nature with wonder and awe
- the ecological-scientific type, which has a need to investigate nature systematically
- the aesthetic type, which considers the beauty of nature to be paramount
- the symbolic type, which conceives nature as a source of symbols and as a means for improving communication
- the humanistic type, which feels emotionally connected with nature, particularly with large vertebrates
- the moralistic type, which feels responsible for nature
- the domineering type, which wants to master the natural world completely
- the negative type, which views nature with fear and aversion

In this typology the concepts of biophobia and biophilia are clearly apparent. Biophobia is the tendency for people to be afraid of animals and natural states and to want to avoid them, as these have, in the past, threatened humans (Ulrich 1993). Biophobia is thought to have a genetic basis and is assigned by Kellert to the negative type. Biophilia, which is the complement to biophobia, is the tendency for people to prefer nature and the natural environment to the human-made, i.e. constructed, environment (Wilson 1993). According to Wilson, biophilia is a genetic adaptation to natural environments that gives humans better survival chances. Kellert's (1980) naturalistic type provides a particularly clear description of this biophile attitude. These two tendencies probably also play important roles in describing other attitudes toward nature.

In addition to this typology, which describes general attitudes towards nature, this literature survey focuses mostly on empirical studies that have considered wilderness as a special kind of nature. These can be divided into studies that (1) record general attitudes to wilderness and phenomena having to do with the spread of wilderness, and (2) analyse people's motives for using wilderness areas and what they need and expect to find there.

Many of the studies we refer to here have been carried out in the United States where, it should be noted, "wilderness" has had a different history to that of "wilderness" in Europe. In Europe, people have tended to understand "wilderness" to mean either fallow land or wasteland, which fitted in with the classical challenge of making "wilderness" useful for humans (Trommer 1997). On the continent of America, European settlers' partial conquest of the wild took place at the same time as industrialisation, and as a result of urban nostalgia, "wilderness" came to represent a space in marked contrast with the urban, civilized world (Stremlow and Sidler 2002). Since the American Revolution, "wilderness" has come to be associated with the notion of unity with nature.

The first type of study, in which people's general attitudes towards wilderness are recorded independently of whether or not they are visitors to a particular area, is rare. Lutz et al. (1999) found, in a questionnaire survey carried out in Canada, that most respondents from both urban and rural areas favoured protecting wilderness areas and viewed wilderness quite positively. But when they were asked to evaluate pictures of landscapes according to their "level of wilderness", it became clear that town and country dwellers had rather different conceptions of wilderness. Rural respondents graded the pictures as much less "wild" than the urban respondents did.

There have also been only a few studies of attitudes regarding previously cultivated or tended areas becoming wild. This may be because the spread of wilderness is a relatively new phenomenon that has only recently begun to have an impact as land use for agriculture and forestry has declined. Studies that focus on the spontaneous reforestation of derelict land are relevant here as this is an important aspect of secondary wilderness. Hunziker (1995, 2000) came to the conclusion that people prefer some natural reforestation on fallow land to having it all traditionally cultivated, but at the same time, they do not like to see reforestation covering extensive areas.

Several studies that focus on people's motives for using wilderness and that were carried out in designated wilderness areas have come up with two rather similar findings: When people join wilderness excursions, the social aspect of being in a group is apparently not an important factor in motivating their participation. On the other hand, the aesthetic quality of a wilderness adventure, i.e. out in the wild, is the most important motive for visiting a wilderness area (Rossman and Ulehla 1977; Brown and Haas 1980; Borrie and Roggenbuck 2001).

Besides the research on people's motives for visiting wilderness areas, a variety of investigations have also been carried out to discover the profiles of the users of these areas in order to adapt the wilderness better to their needs. The aspect that generates the most satisfaction, according to Borrie and Roggenbuck (2001), is having the opportunity to observe wild animals. What people find dissatisfying, on the other hand, is the presence of rubbish, damaged trees and many other visitors.

Cole (2001) compared visitors on day trips with those who stayed overnight. He found no differences between the two groups in how they evaluated wilderness and wilderness areas. The two groups had had comparable previous experience in such areas. Day-trippers, however, seemed less interested in spending time in the wilderness and seemed to make less use of wilderness areas in organising their leisure activities. Other studies have evaluated the places where visitors can stay overnight in wilderness areas to determine what perceptible influence they have on the surrounding vegetation (Farrell et al. 2001). The studies referred to here were carried out in the United States, which means they cannot be compared directly with developments in central Europe where both the extent and history of wilderness areas are very different.

In conclusion, there are many studies of people's experience with nature in designated wilderness areas, but only a few studies of attitudes to the spread of wilderness into previously cultivated and exploited areas or of users' wishes in the rather small areas in which wilderness has spread in Europe.

Goals of the study, questioning techniques and methods

It is forecast that, in the coming years, agriculture and forestry throughout Europe will decline further (Eissing 2002). This means that many areas that are exploited at present will no longer be used and parts of these areas will be taken over by wilderness. Since experts cannot agree on what consequences this development will have, for example, for species diversity, we believe it is important to find out what kinds of attitudes the general public have to wilderness spread and to a passive form of nature conservation with a "wait and see" approach and no direct interventions. We explore the questions: What is the reasoning behind the different stances taken on wilderness and its spread, and what points of view are involved in evaluations of wilderness spread? In addition, we are trying to find out what kinds of perceptions of wilderness are widespread, what the general public want from wilderness areas and what the implications of the findings are for selecting new wilderness conservation areas.

Since there have been only a few studies of people's attitudes to the spread of wilderness, we chose a two-phase research design, consisting of an initial inductive phase that was used as the basis for the following deductive phase. The methodological goal of the exploratory inductive project phase was to develop hypotheses to be tested in a representative survey in Switzerland. The following section describes this exploratory data collection and generation phase, which was based on qualitative interviews. Further below, we describe the procedure used in the representative questionnaire survey.

The inductive-exploratory phase

Area investigated and sampling universe

We chose the wilderness areas in those regions in the German part of Switzerland where wilderness and its spread are the subjects of intense debate. By "wilderness areas" we mean areas in which there has been hardly any human intervention in the form of tending or cultivation or in which hardly any such measures are still being implemented. The study areas were: (1) a forest near Zürich that had previously been commercially exploited, (2) a re-naturalised floodplain area close to an urban area, (3) an overgrown cultivated area, and (4) an area with an Alpine wilderness that is being considered as a possible second Swiss national park. All the selected wilderness areas can still be used without any significant restrictions as recreation areas.

Selection of interviewees

Interviewees were selected from the local population in the regions around the wilderness areas according to the theoretical-sampling method (Strauss 1991) in order to identify as contrastive views as possible (Hunziker 2000). The aim was not to select positions that are representative in a quantitative sense, but rather to have the greatest possible differences between positions. Thus interviewees were chosen to ensure that as wide a range of reasons as possible for opinions on the spread of wilderness would be covered. This selection procedure meant that all thematically relevant positions in the sampling universe could be identified and then taken into account in the sampling procedure (Fig. 1).

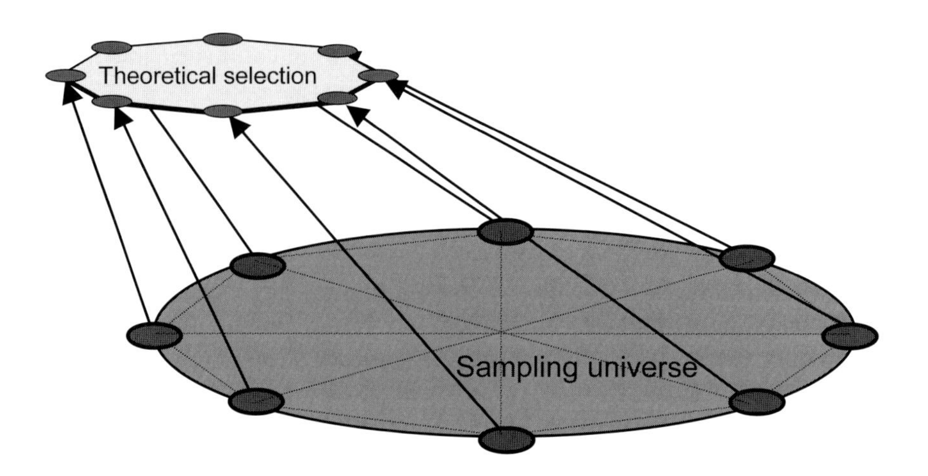

Fig. 1. Theoretical-sampling according to Strauss (1991). The *small dark circles* represent the extreme positions in the sampling universe that have to be considered in the theoretical selection (figure adapted from Hunziker 2000).

The first step was to prepare a hypothetical sample that would be relevant to the study, based on the literature on related areas. For this we drew on studies of attitudes to the re-introduction of wild predators (Egli et al. 2001) and of society's requirements for forest areas (BUWAL 1999). This research has identified some of the demographic characteristics that are closely associated with different attitudes to these phenomena. These characteristics, in this case the age of the interviewees and their professional relationship with nature, are therefore likely to be relevant to research on attitudes to the spread of wilderness (Table 1).

Data collection

Problem-centred interviews were used for the data collection (Witzel 1989). These followed guidelines which suggested potential questions about the following themes:

- experience with nature
- time spent "in nature"
- importance of the natural world during childhood
- type of relationship with nature
- awareness and evaluation of changes in nature during recent years in the region where the interviewee lives

- desired landscape developments

The questions could be used as narration prompts if necessary, but the interviewee determined how the talk developed and the order in which the themes were discussed.

The expressions "wilderness" and "wilderness spread" were avoided throughout the interviews as they may have negative connotations (Stremlow and Sidler 2002). Instead phrases such as "not influencing nature", "having less influence", "leaving nature more to its own devices", and so on were used.

Each interview lasted between 90 and 120 minutes. They were all recorded and later transcribed.

Table 1. Demographic data on the interviewees

No.	Age	Sex	Occupation
1	30	M	Biologist
2	51	M	Management consultant
3	49	M	Forester
4	52	M	Farmer
5	32	F	Clerical worker
6	46	M	Engineer
7	52	M	Lawyer
8	82	M	Town clerk
9	56	F	Housewife
10	67	F	Careers officer
11	26	F	Student
12	23	M	Car mechanic
13	39	F	Bank businesswoman
14	69	F	Teacher
15	31	F	Graphic artist

Average age of the interviewees = 47, standard deviation = 17.13 years. Distribution according to sex: 7 female and 8 male interviewees.

Analysis of the interviews

All the interviews were independently evaluated by two people according to Grounded Theory (Glaser and Strauss 1998).

(1) The transcripts were analysed line by line to identify any comments expressing attitudes about the spread of wilderness and the reasons given for these attitudes. These parts of the text were then given brief labels to summarise the contents. The resulting codes were considered in context, i.e. they were compared with other statements in the interviews. Similar

statements were then grouped into categories. Some of the findings from the evaluation of the interviews were used to add themes to the guidelines if they had not already been documented or to give them more weight in subsequent interviews.

(2) In the intrapersonal case reconstruction, the arguments of the individual interviewees were analysed in depth, summarised and cross-checked for logical consistency. We then checked whether the categories already proposed took such arguments adequately into account.

All the steps in the evaluation were carried out for each interview by two people working independently. Only when these steps had been completed did the analysts discuss their results. Both evaluations, however, resulted in each case in a typology with three different views, with the two typologies matching quite well. The underlying categories in the two evaluations were also largely identical.

At the end of the evaluation process, a short list of key categories was drawn up and a typology proposed. This typology described the most important positions in the relationship between people, wilderness, and wilderness spread, with the "typical" grouping of the categories and the category characteristics.

The deductive phase

The inductive-explorative phase led to the formulation of numerous hypotheses and questions that could then be explored in the representative questionnaire survey in Switzerland.

The questionnaire

In the multiple-choice questionnaire respondents were asked, among other things, how often they spent time outdoors in "nature", what they did outdoors, and what their attitudes toward nature and landscape, as well as toward the spread of wilderness were. In addition, they were asked about how they defined wilderness both in general terms and in connection with different types of natural landscapes. There were also questions about what they thought the most important characteristics of wilderness areas were, what they would like these areas to look like and how they would like them to be managed. The most important demographic data were recorded at the end of the questionnaire.

The random sample

The questionnaire was sent to 4,000 randomly selected households throughout Switzerland. The first person over the age of sixteen in the household to have a birthday in the year was asked to complete the questionnaire. This selection procedure was intended to ensure that the questionnaire was completed by the same number of women as men, and that all age groups were equally represented.

Of the 4,000 questionnaires sent out, 182 were undeliverable, and 1,536 were returned completed for an actual response rate of 40.23%. The main demographic data from the survey and the comparable data from the official national statistics are shown in Table 2.

The characteristics of the respondents were found to deviate slightly from the official national statistics. Women, with a response rate of 43.3%, were slightly underrepresented in our survey. The age distribution was quite similar to that of the general population, with only the category of people under the age of 39 years being underrepresented. Foreigners, however, made up only 9% of the respondents, which is considerably less than in the wider population. It is possible that language difficulties could have acted as a barrier here. The questionnaire was distributed in the three official Swiss languages (German, French and Italian), and people who are not fluent in any of these languages would have found it difficult to complete the questionnaire.

Table 2. Comparison of central variables with the official national figures

	Wilderness survey 2002	Statistical yearbook 2000
Age		
Up to 39	30.2%	38.3%
40-64	49.9%	42.5%
65 or older	19.9%	20.2%
Sex		
Female	43.3%	51.2%
Male	56.7%	48.8%
Nationality		
Swiss	90.6%	80.6%
Other	9.4%	19.4%

Findings

Results of the interviews

In our evaluations of the interviews, we were able, as a result of the procedure described above, to isolate the key categories that appeared to have influenced people's attitudes to the spread of wilderness and to new wilderness areas becoming established as well. The categories we identified were: *beauty, contrast, diversity, usefulness, safety, past as a reference point* and *freedom from regulations* (Fig. 2).

The interviews showed that what influences people most in deciding to favour or reject wilderness spread is not whether nature is "managed" or left untouched, but rather their subjective perceptions of beauty. The interviewees varied greatly in their evaluations of beauty. Some thought a forest left uncultivated lost beauty, whereas others found a landscape where humans had had little impact to be more beautiful.

Beauty

The category *beauty* was an important factor for interviewees in deciding whether to view wilderness spread favourably or not. When the interviewees considered diversity and/or contrast in a landscape to be a prerequisite for judging it beautiful, the category *beauty* tended to correlate closely with the categories *diversity* and *contrast.*

Diversity

For some of the respondents, the dimension *diversity* appeared to be a goal to be reached and usually meant species diversity and less often the structural diversity of landscapes. "Giving up" the cultivation of previously tended areas was usually associated with a reduction in species diversity, and thus with a move further away from the desired goal of achieving the greatest possible diversity. A desire for greater species diversity tended, therefore, to be closely associated with a rather negative attitude towards wilderness spread.

No. 9: "...without human intervention we would have only forest, ...because we want nature to be varied, people need to intervene."

Contrast

Some of the respondents, moreover, particularly favoured those landscapes they felt contrasted with their local, everyday landscape.

No. 6: "...it's as if you would walk from silence into Beethoven's Ninth Symphony, isn't it? The contrast is so strong, which is what I like."

Wilderness areas do provide, without doubt, a contrast with cultivated landscapes, but this discrepancy alone does not guarantee positive attitudes to wilderness areas. What is decisive is how people assess a particular landscape aesthetically. Some, as mentioned above, associated the category *contrast* strongly with the perceived *beauty* of areas that had become wild.

Usefulness

The analysis of the interviews showed that most of the interviewees regarded nature in general and areas to which wilderness had spread from a utilitarian point of view. Several stressed above all the recreational function of nature.

No. 11: "...well, nature provides space for recreation and hobbies like hunting or skiing or biking ... I really enjoy biking up a pass ..."

Others, however, consider one of the main functions of nature is ensuring continued human survival — or expressed more generally — they see the economic functions of nature as more important.

No. 4: "...yes, purely from the point of view of profitability, I must say a monoculture is more interesting, isn't it?"

Safety

An important factor for some of the interviewees toward the spread of wilderness was their concern for ensuring the safety of the public. Protecting processes, i.e. protecting dynamic nature, in which unpredictable processes outside of human control take place, was mentioned in connection with threats to human safety.

No. 9: "...yes, well, if old wood is lying around in there and it can start to slide, or also, everything slides down together,... they were just afraid that it might all come down on the road and bury it."

The past as a reference point

Elderly people and those professionally connected with nature tended to consider the landscape of the period when cultivation was intensive as the norm. This preference for the past is not only motivated by considerations of usefulness, but more importantly by reverence and respect for the human effort made by previous generations. It is this aspect that made these interviewees keen to preserve the previous state and to reject secondary wilderness.

No. 8: "…I think it is a disgrace, really, that we should let all this forest, which has been tended for decades or almost centuries and used sensibly, simply to go back to nature. I don't plan to ever get to like this idea."

Freedom from regulations

Several interviewees mentioned that the absence of regulations and duties when they were out in the wild made them predisposed to favour wilderness areas. Most of them were against having rules and regulations restricting behaviour in wilderness areas.

No. 1: "...With large agglomerations I think it is especially important for the people who live there to have access to areas and be able to find such areas and enjoy themselves there with as little management as possible..."

Typology of human-wilderness relationships

Since the values and the weighting of the key categories described differentiate well among three different forms of reasoning, we have developed a typology of human-wilderness relationships (Fig. 2).

The category *beauty* was crucial for all the interviewees in deciding whether to favour or reject the spread of wilderness. How the three groups concretely described what they perceived to be beautiful, however, was very different. This means that, for this category, interviewees' descriptions of the natural landscapes they perceive to be beautiful, i.e. the values of the category *beauty,* are especially important for discriminating among the types.

Type I: Conservative wilderness opponents

The pattern of reasoning associated with this type is made up of the categories *safety*, *past as a reference point* and *usefulness*. People who adopted this style of reasoning view uncultivated or no-longer-cultivated nature as a threat. They reject secondary wilderness as they see it as a potential threat to populated areas. Moreover, the fact that such wilderness areas are not economically useful can be decisive in making them reject nature "gone wild". The *conservative wilderness opponents* feel obliged to continue the efforts of their ancestors and to maintain tended land or make as much land as possible usable. This type of person also perceives cultivated land to be more beautiful than areas left to develop untouched.

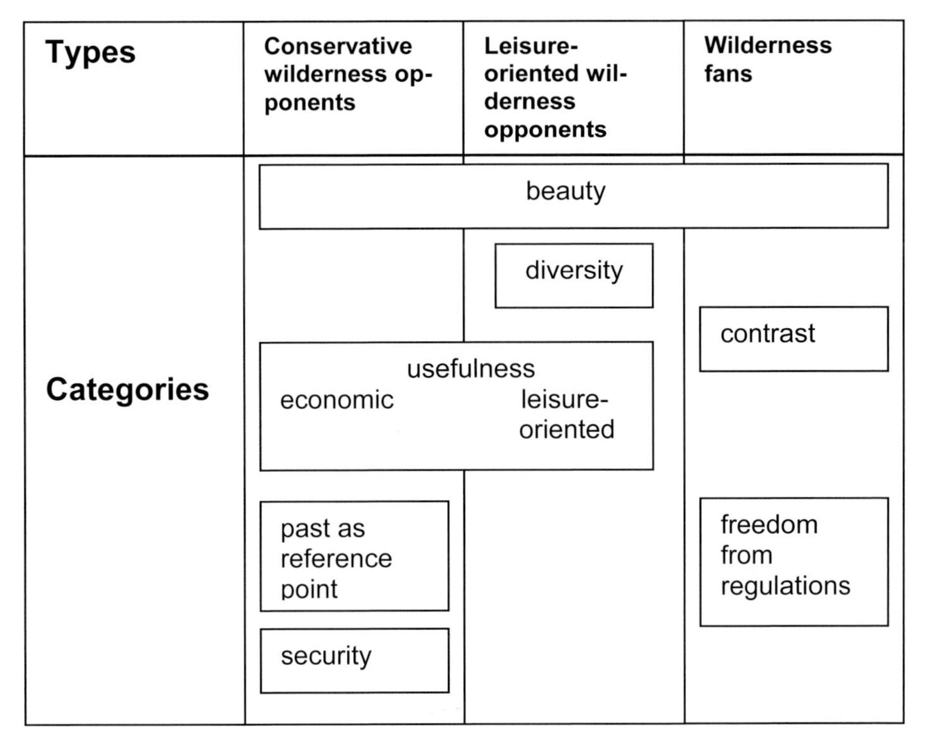

Fig. 2. Typology of the human-wilderness relationship and the attitudinal relevance of the key categories in the human-nature relationship. Beauty, for example, is highly relevant for all types in influencing their attitudes toward wilderness, whereas freedom from regulations was only relevant for the wilderness fans.

Type II: Leisure-oriented wilderness opponents

The categories *diversity* and *usefulness* (in the sense of usefulness for recreation purposes) played a main role in distinguishing the style of reasoning of this type from the others. Their desire to intervene in nature is motivated by their belief that nature needs to be looked after. Planned human interventions care for nature to ensure that it remains varied and ecologically valuable (*diversity*). This type's understanding of nature is in many ways similar to the understanding associated with traditional nature conservation. They are opposed to wilderness and its spread because it is associated with a loss of diversity. They value nature for its usefulness for people and the opportunities it provides for such leisure activities as sport. The *leisure-oriented wilderness opponents* tend to see areas that have gone wild as not very attractive. They perceive nature to be more beautiful if it contains considerable diversity as, for instance, in nature conservation areas where rare species can be observed.

Type III: Wilderness fans

Interviewees belonging to this type tended to see themselves as part of nature and to accept both human influence on nature, so long as this is restricted to populated regions, and the impact of nature and natural events on humans. They are opposed to human intervention to tend and plan nature in the sense of traditional nature conservation. Wilderness is viewed as special primarily because it *contrasts* with inhabited areas formed through human design but also because it is out in the wild that people are free to experience nature without restrictions. Nature to which wilderness has spread is seen as robust enough to withstand human use for recreation without too much damage. Such wild areas are seen by these people as more beautiful than any systems that have been influenced by humans.

Results of the questionnaire survey

The typology described above indicates which criteria and characteristics are involved in evaluations of wilderness and its spread, and what the typical attitudes are and what the reasoning used to justify them is. In the representative survey it became clear that general definitions of wilderness are strongly oriented towards scientific criteria: Most people in the test sample understood "wilderness" to refer to areas still untouched by human influence (90.8%), to abandoned land (64.0%) or to areas with thick vegetation (51.3%). Only 40.9% of the respondents considered renaturalised areas, i.e. areas in which a particular "natural" state is actively restored, to be wilderness (Fig. 3).

Moreover, the majority of respondents (56.4%) thought that, in order for them to feel that they are out in the wild, it was important for there to be large distances between wilderness areas and peri-urban regions.

Only 42.8% could imagine having a wilderness in a peri-urban region, but 52.3% were still in favour of establishing wilderness areas in such regions. This apparently contradictory finding is due to the different criteria used for labelling different types of wilderness: If respondents are asked explicitly which natural spaces in peri-urban regions they feel to be wilderness, their definitions of wilderness change. Derelict land, i.e. plots in peri-urban regions that are no longer used and that have therefore become covered with grass and shrubs, are seen by 66.3% of the respondents to be wilderness areas. Steep ravines, which are largely undisturbed natural areas, parks with flower meadows and shrubs, and near-natural gardens, are mostly not perceived as wilderness.

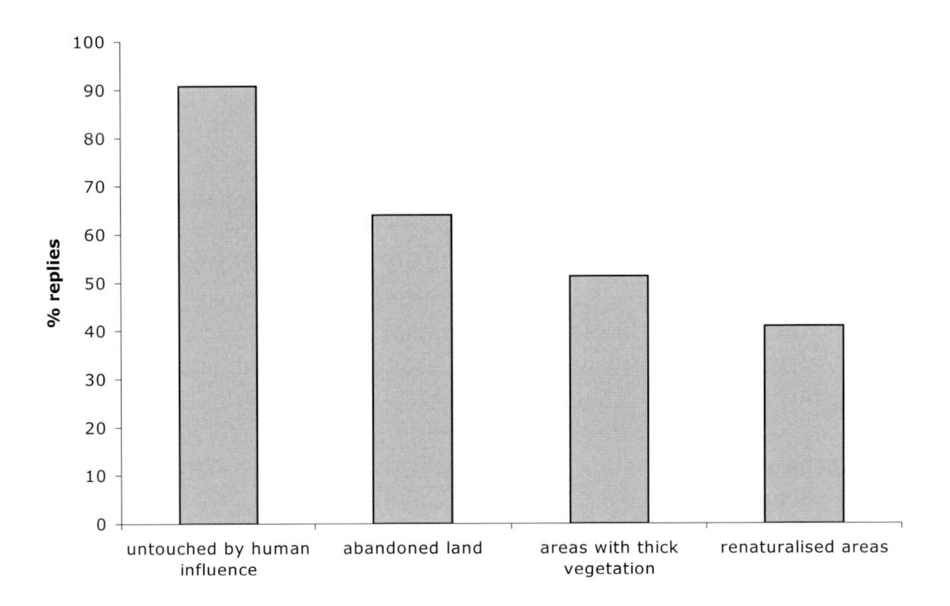

Fig. 3. Criteria of wilderness areas according to the respondents of the questionnaire survey (n=1,536)

The results show that the characteristics of wilderness areas described as "typical" tend to stem from scientific definitions. These characteristics, however, do not necessarily correspond to what people want from wilderness. Typical characteristics of wilderness are seen to be the absence of human influence (68% of replies), the absence of economic exploitation (84.4% of replies) and the absence of use for leisure activities, such as sport (60.3% of replies).

What people want and require in wilderness areas is not, however, reconcilable with these typical characteristics: 91.8% of the respondents want notice boards with information about the local vegetation and 67.8% want a network of paths. Some respondents expect to find even more human features in wilderness areas and to have camp-fire places, benches and rubbish bins (57% of respondents) as well as parking areas for visitors (54.3% of respondents).

Discussion

The results from this qualitative phase have many parallels with Kellert's (1980) typology of human-nature relationships. We would like to mention just a few of these parallels here: The *conservative wilderness opponents* evaluate nature from a utilitarian point of view. They assess the absence of economic exploitation negatively and are opposed to wilderness spread. This kind of view is identical with the utilitarian approach described by Kellert, in which nature is assessed according to how economically useful it is.

The *leisure-oriented wilderness opponents* also tend to have a utilitarian view of nature. Unlike Kellert's (1980) utilitarian type, however, who emphasises the economic usefulness of nature, *leisure-oriented wilderness opponents* tend to value nature most for providing opportunities for leisure activities.

Wilderness fans typically feel they belong to nature and accept natural phenomena unconditionally. In this they correspond to Kellert's naturalistic type, who also expresses a kind of biophilia (Wilson 1993). *Wilderness fans*, however, also maintain that being able to have a break away from the everyday is important, and this is a defining characteristic of a positive attitude to wilderness and wilderness areas. Such a break involves, on the one hand, people being able to contrast the wilderness visually with their everyday landscape and, on the other, having a (desired) absence of rules and regulations in wilderness areas.

How important it is to be free of rules and regulations in these areas is, it seems, one criterion to consider in evaluating wilderness areas. Patterson and Watson's (1998) survey of visitors in wilderness areas also indicates that an important motive for visits was to be able to use the wilderness in an unrestricted and individual way. This aspect of freedom from regulations is closely tied to the potential use of wilderness areas and is thus also indirectly utilitarian.

The numerous parallels between the patterns of reasoning we found and the typologies in Kellert's human-nature relationships suggest that nature involving wilderness and its spread tends to be evaluated according to the same kind of criteria as those for evaluating nature in general. This observation is also supported by Hunziker et al.'s findings (2001). They maintain that people's attitudes to changes in which nature becomes freer and humans have less influence on it, such as the re-introduction of wild animals or spontaneous reforestation, are connected to their attitudes towards the relationship between humans and nature.

Our empirical analysis of attitudes to the spread of wilderness indicates that several aspects of those attitudes apparent in the human-nature relationship are involved. We can conclude that, generally, the spread of wilderness is evaluated according to the same basic criteria as nature in general, but that such evaluations tend to draw on several criteria, including some additional ones reflecting different points of view.

The results of the questionnaire survey suggest that, in addition to a differentiated evaluation of nature and wilderness, people also tend to distinguish different kinds of "wilderness". In defining wilderness in general they tend to refer mostly to assessment factors that are strictly scientific, and wilderness is usually described as completely untouched nature. If wilderness in or near a settled area is considered, however, then other criteria are evoked that have less to do with scientific definitions. Places that have become derelict and that show signs of wilderness spreading are perceived as wilderness, whereas other natural sites that are largely untouched, like steep ravines, are usually not considered as wilderness.

The responses indicate that the omnipresence of human influence in peri-urban areas is incompatible with the notion of untouched nature. Unlike with the general definition of wilderness, for wilderness in a peri-urban region it is not whether nature has been left untouched that is the most important defining criterion, but whether humans have stopped using and influencing it.

The results of this study also suggest there is a discrepancy between people's general perception of existing wilderness and their wishes for what wilderness areas should be like. A defining characteristic of wilderness from the point of view of the majority of the respondents is that it is not useful for people. At the same time, however, these respondents say they want to make it useful for people and to shape it to best fulfill their needs. This is a further indication that the utilitarian view is actually quite widespread. It was also evident in each of the three types distinguished during the qualitative phase.

Conclusions

A sustainable landscape development is only possible if people's preferences are taken into account (Mansvelt and Lubbe 1999). This means allowing for a full range of public opinion when designating places as wilderness areas. The results from the qualitative phase of this study suggest that, in general, the three positions most frequently observed in the human-wilderness relationship differ markedly from each other. Their proponents tend, correspondingly, to have different requirements of wilderness areas. At the same time, the results of both the qualitative phase and the questionnaire survey indicate that the utilitarian view of wilderness is very widespread, but respondents tend to differ considerably in how much they wish wilderness areas to be managed and shaped. The questionnaire survey also indicated that, from the point of view of the public, there is a discrepancy between perceptions of existing areas of wilderness and ideal images of wilderness. This situation means that when areas are designated as wilderness areas, a participatory approach should be taken in which the different stakeholders have opportunities to express their views and interests.

The typology of human-wilderness relationships highlights the motives underlying, or potentially underlying, particular attitudes and the reasoning associated with them. It may be that the wish for wilderness to be planned and designed, which assumes a certain amount of controllability, serves more or less unconsciously to reduce people's fear of the wild.

Since nature takes different forms and the spread of wilderness changes it in different ways, it presents different kinds of threats to different people. Thus there can be no standard solutions for planning wilderness areas. Instead, individual schemes should, in each case, be developed through negotiations with local residents and with the various stakeholders. Our results indicate clearly, however, that the utilitarian point of view is very widespread among the public and should be integrated in plans to designate particular areas as wilderness. One way of addressing these wishes and at the same time doing justice to the original goal of protecting such an area is to set up a system for zoning wilderness areas with (1) key zones, where nature should be allowed to develop freely and largely undisturbed, but which can still be visited on guided tours, and (2) zones designed to cater more to people's need to have space for outdoor activities (e.g. picnicking or practising sports). Only by having schemes in which people can experience wilderness directly will it be possible to increase people's acceptance of wilderness areas and their understanding of natural processes (Schemel 1998). This, in turn, would also positively influence the acceptance, in the long term, of "correct" or "real" wilderness, i.e. of wilderness as defined according to scientific criteria.

Acknowledgements

I would like to thank Bristol Foundation for the funding of the research reported in this article and I am very grateful to Karin Wasem for her assistance in data collection and analysis.

References

Borrie WT, Roggenbuck JW (2001) The dynamic, emergent, and multi-phasic nature of on-site wilderness experiences. Journal of Leisure Research 33:202–228

Broggi M (1999) Ist Wildnis schön und „nützlich"? In: W Konold, R Böcker, U Hampicke (eds) Handbuch Naturschutz und Landschaftspflege: Kompendium zu Schutz und Entwicklung von Lebensräumen und Landschaften (Chap. V-1.1). Ecomed, Landsberg, pp 1–7

Brown PJ, Haas GE (1980) Wilderness recreation experiences: The Rawah Case. Journal of Leisure Research 12:229–241

Bundesamt für Statistik (ed) (1999) Statistisches Jahrbuch der Schweiz 2000. NZZ Verlag, Zürich

BUWAL (1999) Gesellschaftliche Ansprüche an den Schweizer Wald – Meinungsumfrage. Bundesamt für Umwelt, Wald und Landschaft, Bern

Cole DN (2001) Day users in wilderness: How different are they? USDA Forest Service Rocky Mountains Research Station Paper, (RP 31, 9/2001) Odgen, Rocky Mountain Research Station

Egli E, Lüthi B, Hunziker M (2001) Die Akzeptanz des Luchses – Ergebnisse einer Fallstudie im Berner Oberland. Forest, Snow and Landscape Research 76:213–228

Eissing H (2002) Die Wiedergewinnung der Wildnis – Gedanken zu Wildnis und Wildniserfahrung. In: Evangelische Akademie Tutzing (ed) Wildnis vor der Haustür. Morsak, Grafenau, pp 12–24

Farrell T, Hall TE, White DD (2001) Wilderness campers' perception and evaluation of campsite impacts. Journal of Leisure Research 33:229–250

Glaser BG, Strauss AL (1998) Grounded Theory. Strategien qualitativer Forschung. Huber, Bern

Hunziker M (1995) The spontaneous reafforestation in abandoned agricultural lands: perception and aesthetic assessment by locals and tourists. Landscape and Urban Planning 31:399–410

Hunziker M (2000) Einstellungen der Bevölkerung zu möglichen Landschaftsentwicklungen in den Alpen. Eidgenössische Forschungsanstalt WSL, Birmensdorf

Hunziker M, Hoffmann C, Wild S (2001) Die Akzeptanz von Raubtieren, Gründe und Hintergründe – Ergebnisse einer repräsentativen Umfrage in der Schweiz. Forest, Snow and Landscape Research 76:301–326

Kellert SR (1980) Contemporary values of wildlife in American Society. In: Shaw WW Zube EH (eds) Wildlife Values. Center for Assessment of Noncommodity Natural Resource Values (Institutional Series Report Nr. 1). University of Arizona, Tucson, pp 241–267

Lutz AR, Simpson-Housley P, de Man AF (1999) Wilderness — Rural and urban attitudes and perceptions. Environment and Behavior 31:259–266

Mansvelt v JD, Lubbe vd MJ (1999) Checklist for sustainable landscape management. Final report of the EU concerted action AIR3-CT93-1210: The land-

scape and nature production capacity of organic/sustainable types of agriculture. Elsevier, Amsterdam
Patterson ME, Watson AE (1998) A hermeneutic approach to studying the nature of wilderness experiences. Journal of Leisure Research 30:423–452
Rossmann BB, Ulehla ZJ (1977) Psychological reward values associated with wilderness use. A functional-reinforcement approach. Environment and Behavior 9:41–66
Schemel H-J (1998) Naturerfahrungsräume. Bundesamt für Naturschutz, Bonn-Bad Godesberg
Strauss AL (1991) Grundlagen qualitativer Sozialforschung. Wilhelm Fink Verlag, München
Stremlow M, Sidler C (2002) Schreibzüge durch die Wildnis. Wildnisvorstellungen in Literatur und Printmedien der Schweiz. Haupt, Bern
Trommer G (1997) Wilderness, Wildnis oder Verwilderung – Was können und was sollen wir wollen? In: Bayerische Akademie für Naturschutz und Landschaftspflege (eds) Wildnis – ein neues Leitbild? Möglichkeiten und Grenzen ungestörter Naturentwicklung für Mitteleuropa. Laufener Seminarbeiträge 1/97. Grauer, Laufen/Salzach, pp 5–8
Ulrich RS (1993) Biophilia, biophobia, and natural landscapes. In: Kellert SR, Wilson EO (eds) The biophilia hypothesis. Island Press, Washington DC, pp 73–137
Wilson EO (1993) Biophilia and the conservation ethic. In: Kellert SR, Wilson EO (eds) The biophilia hypothesis. Island Press, Washington DC, pp 31–41
Witzel A (1989) Das problemzentrierte Interview. In: Jüttemann G (ed) Qualitative Forschung in der Psychologie – Grundfragen, Verfahrensweisen, Anwendungsfelder. Beltz, Weinheim, pp 227–255

Surrogate Nature or Wilderness? Social Perceptions and Notions of Nature in an Urban Context

Dieter Rink
in cooperation with Rico Emmrich

Department of Economy, Sociology and Law, UFZ Centre for Environmental Research Leipzig-Halle

Beautifully wild? A discussion of a wilderness in the city

"Wilderness" has become a catchword in the current debate on urban development. Recently the subject has been discovered by urban planners in connection with urban restructuring in eastern Germany. As it seems unlikely that many of the sites are to be redeveloped even in the longer term, there is a need for other uses. In addition to lawns, new parks and gardens, the idea of an urban wilderness on one's doorstep as an area for experiencing nature has come under discussion. This would enhance the areas concerned as well as being cost-effective. Is urban wilderness, therefore, a smart and inexpensive planning idea?

Conservationists have also taken hold of the subject of "wilderness in the city" (see Hard and Kruckemeyer 1993; Erdmann 2002; Wächter 2003). The wild side of nature is to be brought closer to the city dweller. Educational packages especially for children and young people are aimed at developing and strengthening attachments to nature and environmental awareness (see Nymphius and Trust 2001). In addition, "urban wilderness" is linked to the hope that people will eventually be persuaded against taking long journeys to experience nature and wilderness. Urban wilderness appears to extend opportunities for nature conservation into the urban environment and to establish links, for example, with teaching and discussions of sustainability. Can the wilderness concept increase the acceptance of nature conservation in the urban environment?

The term "wilderness" has had an incredible public impact in recent years, from about the middle of the 1990s. Because environmental problems were at the centre of public discussion in the 1970s and 1980s, the

Kowarik I, Körner S (eds) Wild Urban Woodlands.

terms nature and wilderness shifted more strongly into the centre. While "environment", with all its associated problems, has negative connotations (dirt, pollution, destruction), "nature" and especially "wilderness" are presented as unpolluted terms with positive connotations or are perceived as such in the public awareness (Kuckartz 2002).

Especially in the context of the city, however, a few questions arise: Does the term "wilderness" work in the city—is it seen here as something positive? As the environment becomes overgrown, will it be seen as an enhancement or as a sign of decline and decay? Is urban wilderness attractive? And finally, will it be regarded as nature that is worth protecting?

When ideas of usefulness and protection are more closely investigated, it becomes clear that not only should "the" urban nature be differentiated, but the city's inhabitants should be as well. Who particularly values urban nature and why? What models/images of urban nature exist among various social groups and how are they connected to rights of use and protection? And do these ideas, wishes and acquired perceptions relate to nature conservation in the city?

We explored these questions of the perception and valuation of urban wilderness in focus-group discussions in a research project at the UFZ.

Empirical results of social perceptions and notions of nature in an urban context

In the group discussions with selected groups, ideas and perceptions of urban nature and links with questions of nature conservation in the city and the use of urban nature were considered (for details see Rink 2003; for a more general discussion of the idea of nature among social groups see Rink 2002; Brand et al. 2003). Questions such as the following were asked: What comes to mind when you hear the term "urban nature"? What does it cover? How important is urban nature for you compared to other types of nature (ranking)? How do you experience urban nature? How do you use urban nature in your everyday life? Is it important to protect urban nature?

The groups were each presented with forms of urban nature which they then had to rank in terms of attractiveness and how deserving the areas were of protection (see Fig. 1). The vertical axis indicates the level of attractiveness (hierarchically ranked), the horizontal axis the degree of naturalness. Brownfield nature was assessed on the basis of photos: the groups were each shown two photos of brownfield nature (a derelict industrial site and an overgrown railway shed). The same concepts were used for a second task, in which the interviewees were requested to rank different forms of urban nature by the degree to which they are worth preserving (see Fig. 2). Both figures are taken from group interview G1.

Group discussions were held with 5 groups: university students (7 members of a flat-sharing community), mothers of young children (6), school children (15 members of an 8th grade class), dog owners (5) and garden-allotment holders (9) between November 2002 and May 2003. The groups were selected such that the members knew and were in regular contact with each other. The groups selected were also expected to make greater use of urban nature than other groups.

These discussions were recorded in full on tape and then evaluated on a topic-related basis. The essential comments on questions relating to the main thread of the argument were interpreted, collated and condensed in several stages.

What is perceived as urban nature?

This was the central question of the group discussions, which ran through the discussions like a leitmotif. The following ideas have been compressed from the wide range of responses and represent perceptions of urban nature:

1. Everything green in the city

... the Auwald (in general "lowland forest", but more specifically the "Leipzig Auwald"), parks, streets—not only green but anything alive
... fairly large green areas, integrated systems, nothing isolated

2. Structured and tended nature

- "garden nature": distinguished from nature that is not structured or tended, i.e. spontaneous or ruderal nature
- created nature or "artificial nature": in contrast with untouched nature, or nature that is left to itself
- "useless" or "social" nature: distinction from rural nature that is used economically and has been characterised as such for centuries

"Urban nature is also structured ... primulas out, primulas in ... and look after all of it, try to keep everything short and so on ... and so these little parks, these green spaces, they always try to keep them under control just because it's part of the city image."

3. Communal nature

- nature for which the local authorities bear responsibility (in contrast with private nature or commercial nature)

- nature which is there for the public and which is made for them (in contrast with greenery that is not accessible)
- nature that is geared to particular purposes, such as for recreation, sport, games (in contrast with nature that is used commercially or that is of no direct human benefit)
- privacy is important, lack of privacy is mentioned as a primary argument for non-use (e.g. "lying around, sunbathing")

4. Everyday nature

- in contrast with excursion and holiday nature (which is superior or exotic)
- nature that one perceives in everyday contexts and uses (paths, sports grounds, what one finds while going for a stroll, taking the dog out, etc.)
- nature of which one does not make any high or excessive demands

5. Surrogate nature

- not real, "natural" nature (too small, you can hear or see the city, too structured or subject to human influence)
- restricted opportunities for use
- low experiential value
- low value in terms of form (monotonous, "measured off", "from the drawing board")

"Anyway I only find it a kind of surrogate satisfaction. I regularly escape to the park to get some peace but when I am really out in the country it's quite different."

6. Variety of natural forms

- structured and unstructured, artistically staged
- small areas and large areas
- a great deal of variety in a small space
- a variety of forms and partly exotic
- a variety of possible uses

In the course of the group discussions, the participants were asked to evaluate various forms of urban nature based on particular attributes and arrange them accordingly. The qualities by which the forms of urban nature were to be evaluated were the perceived attractiveness or non-attractiveness and the naturalness or artificiality of the urban nature form (see Fig. 1).

The Auwald (a riparian forest) in particular is estimated as being particularly attractive and natural. Botanical gardens and municipal parks are perceived as being more artificial and structured, but still attractive. On the other hand, fallow areas and industrial areas that have been grassed over are seen as unattractive and natural. Urban parks, green squares and green spaces are perceived as being in the category of the non-attractive and artificial. The perceived attractiveness of urban nature is accordingly measured more by its naturalness.

If one included verbal statements about other topics it can be said that the following are regarded as characteristics of *attractive urban nature*:

- it is free of other uses, which prevent or inhibit its use (housing, traffic)
- it is distinguished by a certain expanse
- it is undisturbed: "You're alone there", it is peaceful
- one is screened off and you can "be yourself" (and are not clearly exposed to glances from other blocks of houses)
- it was not designed exclusively for certain purposes but (also) left as it was found (no trails, signs, climbing aids etc.) and which you can therefore also
- use without drastic restrictions

When asked about attractive forms of nature the people questioned indicated the following associations: mountain landscape, landscape where settlements are as far away as possible, and where one can wander undisturbed: "Where it is free of people, where you almost feel ill again because of all the creepy-crawlies and where you can't shelter in the nearest pub when it rains or the nearest bush hut." Coastal landscape, Auwald, peace, where there's nobody.

Some answers appeared to suggest a dilemma between ideal perceptions of nature and the landscapes and forms of nature that actually occur: "I find it difficult. I am a bit fussy. For me to like it, it mustn't be too high, too cold or too hot. There mustn't be too many insects. I don't think it's possible for there to be a landscape that I find really beautiful."

People obviously want something idyllic with the best possible weather, not too hot but with plenty of sun, and no risk of skin cancer; turquoise-coloured water but without any jellyfish or crabs that you might tread on by mistake; no people; a small palm tree bending towards the water; and a hut right on the beach, but again with the necessary luxury of course.

This might give urban nature the opportunity to function as "surrogate nature", as an alternative for unrealisable desires, in line with the maxim: before I let myself be disillusioned by my unreasonable expectations, I'd sooner go to the botanical garden or simply lie down in the municipal park.

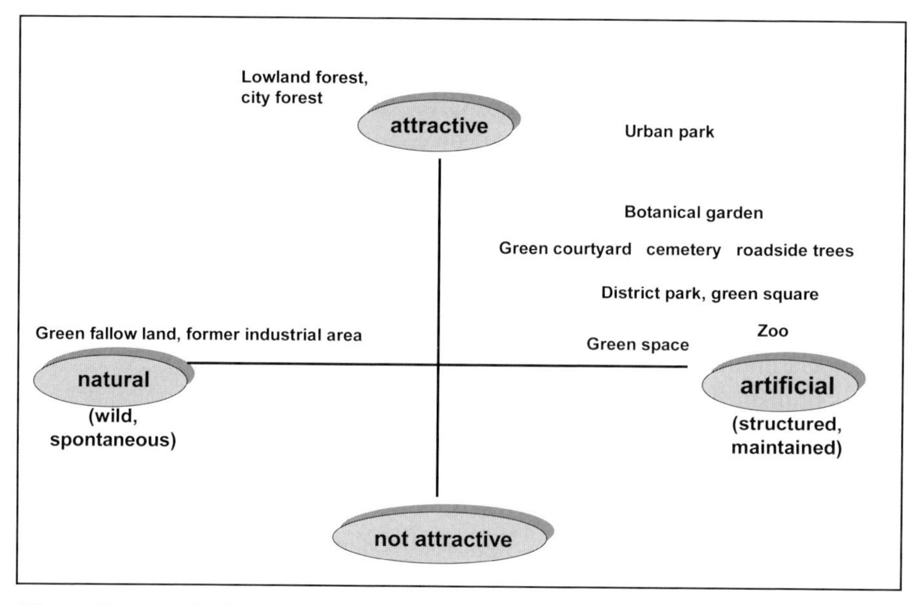

Fig. 1. Forms of urban nature evaluated in terms of their attractiveness and naturalness

There were no systematic questions on the *use* of urban nature. Nevertheless, there were some indications of the use of municipal greenery in the comments of the interviewees. Urban nature is used to a greater or lesser extent on an everyday basis—depending on social group, age and marital status—and so the Auwald is used as a pleasant and safe route to places of work and study, but not as a place that one consciously seeks out for a part of one's leisure. Although urban nature provides relaxation and rest, this is not regular and is rarely used in a purposeful way. It does not therefore have an independent value as recreation (such as by comparison with a recreational area in the nature of the surrounding countryside or of a holiday). This corresponds to the fact that urban nature does not have a high value attached to it compared with "true" nature or, as a participant in a discussion described it, "urban nature is surrogate".

Protecting urban nature?

The participants in the group discussions were asked to evaluate various forms of urban nature in terms of how worthy they were of protection and to prioritise them (see Fig. 2). The participants answered that there should be nature in the city and it should also be protected (from development, commercial use). The large integrated parks in particular, the Auwald etc. should be retained. But roadside trees, yards covered with greenery and allotments should not be sacrificed for arterial roads either. Active personal contribution was ruled out (in the form of tree-sponsoring, horticultural care for a garden, helping to create a park, playground, inner courtyard etc.):

- people would only commit themselves to protecting fairly large parks or urban forests, and not for every little park or every roadside tree (e.g. in citizens' action groups, petitions)
- collecting an additional financial contribution or charging fees (e.g. entrance to a park) was rejected

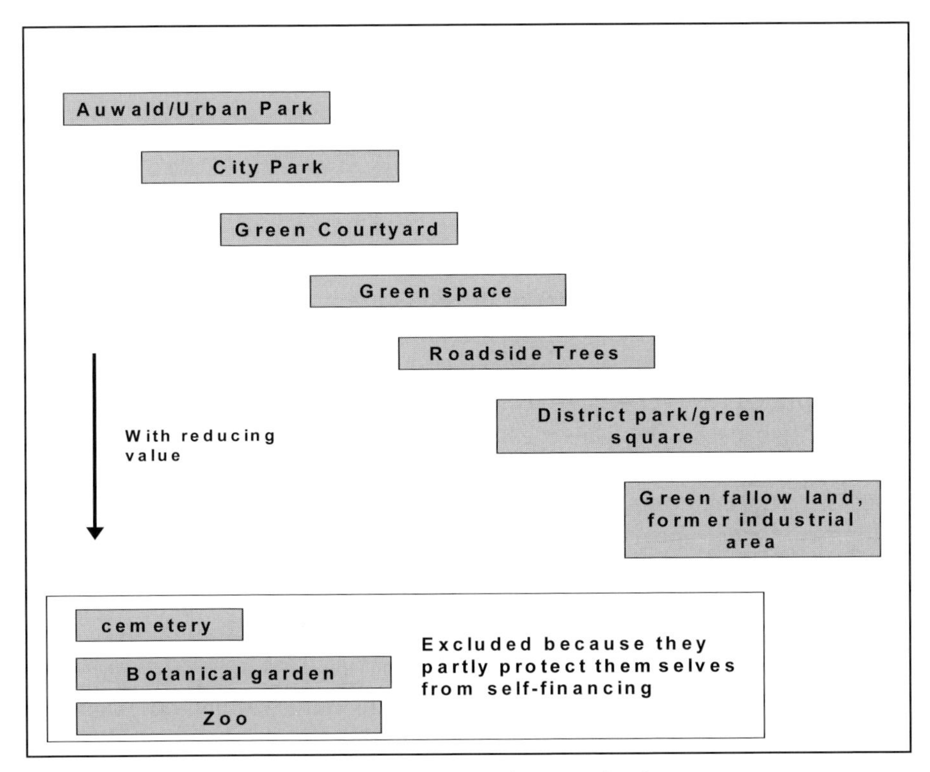

Fig. 2. Which forms of urban nature are worth protecting?

What criteria should one apply when deciding which natural urban forms are particularly worthy of protection and which not? If one considers Fig. 2 on the hierarchy of protection for urban nature, a few parallels arise with regard to the various forms of urban nature presented above as far as their assessment in terms of attractiveness and naturalness is concerned. Forms of urban nature categorised here as being particularly worthy of protection are in general also those to which a more or less high level of attractiveness is ascribed (the Auwald, municipal parks and yards covered with greenery).

Why should nature in the city be protected? Urban nature is definitely regarded as something valuable (e.g. the municipal park), but not from an ecologically motivated position or from a (primary) perspective of nature conservation. Uniqueness, naturalness or diversity of species are not motifs that people put forward in support of nature conservation in the city. And if they do (as for example with the Auwald), then this is least of all true for spontaneous or fallow nature.

"Fresh air", "hygiene", "health" are catchwords that are used to plead for the protection of urban nature. The positive impact on the urban landscape, i.e. aesthetic perceptions of variety, diversity and recreation are other factors justifying protection. Sweeping motifs such as "all nature is worth protecting" were also heard in the discussions.

Wilderness in the city

A special and contentious form of urban nature is represented by so-called spontaneous or ruderal vegetation, as can be found in abandoned areas, for example. There have been urban abandoned areas and spontaneous vegetation, in principle, ever since there have been cities, briefly as an interval between two uses. They have existed more extensively and longer since World War II or de-industrialisation processes. More recently they have also arisen in the course of demolition and rebuilding during urban redevelopment. Although their value was discovered in connection with the environmental debate in the 1970s, these areas were not stylised as "(urban) wilderness" until the last few years.

The "Wilderness in Germany" and "Wilderness in Berlin" campaigns of the German conservation association BUND have become especially familiar. Here "wilderness" is defined as "leaving nature alone, not planning, not interfering in a controlling fashion, but allowing to develop". Wilderness is a process that requires space and time. The reasons given for its being necessary are eco-centric: "We need wilderness as nature's experimental area for the coming and going of species". Evolution as the basis of all life requires freedom (see: http://www.wildnis-in-berlin.de/wildnis.html). At the same time however at least a reference to the social benefits is attempted. "For a high quality of life it (is) important to have ... a juxtaposition of urban and cultivated landscape and free natural development."

Meanwhile, however, the hope shone through that the "wilderness at the doorstep" would persuade people to refrain from going on "long journeys to experience nature and wilderness" (ibid.). With urban wilderness, the attempt was initially made to develop a semantic competition with genuine wilderness—with the jungle and the primeval forests. Urban wilderness was to symbolise the antithesis of the cultivated landscape or the landscape subject to extensive commercial use as well as to serve as the opposite of a sterile green space.

In the group discussions, people were explicitly asked about their assessment of spontaneous manifestations of nature in the city. This was based on the hypothesis that the majority of those questioned tended to view these urban forms of nature either sceptically or negatively. What ideas those questioned have of wilderness and whether they could connect these ideas with a "wilderness in the city" was another topic of discussion.

Fallow and ruderal vegetation was not mentioned spontaneously in answer to the question of what urban nature is, but was only included hesitatingly after further enquiry. It was not necessarily ascribed to urban nature, because it is not structured and cared for and serves no purpose, and because it cannot be used. It is predominantly felt to be wild with weeds. It is associated with dirt and rubbish as well as danger; it is linked with fear and reference is made to the risk of injury. This nature is only seen as being of value for children—as locations for adventure playgrounds. This kind of urban nature is not felt to be wilderness, on enquiry this idea was firmly rejected. Wilderness is another nature, not to be found in the city.

"Well, wilderness for me I think is where there's no trace of a city or a village or anything created by human beings for mile after mile. So I don't think I could find any wilderness in the Auwald, because the noises that you hear still tell you that you're in the middle of the city."

The group of young people (15 girls and boys from an 8th grade class) nevertheless made a more positive assessment of fallow nature than all the other groups. They said they spent a lot of time there and liked to play there.

To record the perceptions and notions that the interviewees had with regard to fallow nature, they were shown two pictures of places that had run wild. The responses revealed a variety of judgments: people find spontaneous forms of vegetation quite good if everything is overgrown from lack of use and nature has reconquered a piece of land. This is judged as being completely "aesthetic", but is at the same time associated with certain reservations. Firstly these abandoned areas or places of spontaneous nature must not be combined with dirt, rubbish and mess.

"What I actually find OK are old industrial sites, when they're not used and gradually get overgrown I don't think that's at all bad and they may even look quite nice. What I don't like is when you get piles of rubbish building up, like little waste dumps."

Secondly, one must also be able to use these areas. Minimal structure should be provided.

"The problem, I think, would be like in the park if you let it get overgrown, I don't know if it's still usable... You get the stinging nettles first, then the thistles. And if a park's just a lot of stinging nettles and thistles I don't think anybody's going to use it any more."

Thirdly the context of the fallow nature is crucial. Can one see any intention behind it? Is it embedded in a nature conservation strategy, for example? Does it produce any other interpretations and assessments? If people perceive these fallow areas or places of spontaneous nature as something consciously intended and known to have been left alone, they will accept them more readily. Other associations will then arise: the image will not be dominated by the interruption of use or the lack of maintenance.

"I just don't believe that it's been left like that intentionally. There simply isn't enough money for it."

"And if it is undefined, it will soon become a tip and an eyesore, but actually it's only growing." Rubbish will collect at these places, "because many people think, they're not doing anything, we can dump what we like."

In fact, people can realise that somebody has taken pains about a place, that there is an idea behind it, that it was intended and planned and structured in line with this idea.

"If, for example, here at ... Square everything was allowed to grow as it wanted, I think that would get to be a problem. Everybody would feel 'That's really not on'. Because it is supposed to be a park and people have a certain idea in their heads as to what a park should look like, and the image doesn't match and then some weeds or other took over and nobody bothered, then we'd all whinge about it. Although people might say they would like it if this piece of land went back to nature, there it's not intended and that's why I don't think it's okay."

This gives rise to a possible communication strategy for coping with abandoned areas and land with spontaneous nature, which the participants also spelt out:

"Well, with a park, [spontaneous vegetation] would only be accepted if there was a heavily publicised project with the press and all this park-going-back-to-nature stuff, then everybody would suddenly find it really cool."

"I think that the definition is quite important, that people want to know what it's all about. If someplace other simply grows and it's something indefinable, what can you use it for? People always want a use, so that you can define it as a park so the children can play there or as an adventure playground."

The results of the group discussions not only suggest that we should differentiate between different kinds of urban nature, but also between different forms of fallow nature and spontaneous nature. The size, usefulness, aesthetics and the location are possible criteria by which one might order and analyse various forms of spontaneous vegetation in the city. This could also be a basis for communication. Such a reassignment of fallow nature as wilderness, wild greenery or urban wilderness certainly involves complex and, in particular, long-term processes, which do not follow any simple chain of cause and effect. This should be borne in mind when one considers the feasibility and effects of such public awareness campaigns.

Lovely and wild?

The demands placed on urban nature – as was made clear by the questioning – are different from those placed on nature outside of the city. Urban nature is its own form of nature, which cannot compete with "nature outside". It makes its own demands and is placed in an urban context. Its special character for people is derived from the large and growing variety of its forms and the resultant variety of individual and group-specific uses. On the other hand, most city-dwellers are hardly aware of the species diversity of urban nature and this idea is barely significant as far as use is concerned.

It must be emphasised that ideas have become established as to what urban nature (such as in the garden, the park or a green space) should look like. This relates both to the form it takes (in Germany perhaps the English tradition of romantic parks and gardens), and their maintenance (closely cut lawns in green spaces). The fact that the attractiveness of an urban form of nature cannot be exclusively derived from its naturalness or artificiality has been demonstrated. Other possible determining factors, which influence the perceived attractiveness of urban natural forms, might be the size and extent of the natural form, the context (in the sense of neighbourhood, localities and anything else that contributes to the definition) and usefulness. It is still somewhat unclear what the determining factors of an attractive urban natural form are. We would have to investigate further to determine to what extent the specific lifestyles of those using these spaces affect the combination of determining factors referred to.

Nature conservation in the city does not meet with total rejection in the urban population; indeed it can rely on a broad general acceptance. It will mainly be more difficult in individual instances, where there are conflicts of interest in terms of use. In the city, nature conservation has even greater difficulty in claiming a special value than "outside". The reference point for evaluating whether urban nature is worth protecting is not its ecological function but its particular structure, its symbolic function, its use or benefits and its relevance for human health and well-being. A process of protection in the city runs up not only against ideas of order, cleanliness and maintenance but also against pronounced ideas and images of what nature in the city should look like. Last but not least, nature in the city must not disturb the city dwellers, it must not smell, nor must it be a refuge for tiresome insects. It should please stay put and do what was planned for it! It takes a long time for such ideas and images to change radically, and they are much less susceptible to ecological arguments than they are to genuinely aesthetic ones.

In this study, a close correlation became clear between the structure of urban nature and ideas about what is worth protecting. Clearly it is the images of parks and gardens that have been handed down for generations, which characterise patterns of perception and evaluation and which wilderness in the city fails to match. The "wilderness" concept does not therefore appear very suitable for promoting new design options and communicating ideas about nature conservation in the city. This will be obvious in attempts to stylise small fallow areas, path and road boundaries and little wooded areas as city wilderness. There is likely to be a greater risk of the city planners and nature conservationists making themselves ridiculous than of increasing acceptance of the new greenery. To stylise spontaneous urban nature as a wilderness doesn't work because wilderness is linked with quite different associations. If attempts are made by nature-conservation campaigns to relate these areas to forms of urban nature, the effect may be counter-productive because the deficient character of urban nature will be even more strongly emphasised as a result.

It became clear in the discussions that one must also be able to use the "wilderness"; the use must be promoted by some sort of structure, however minimal. Shutting out people completely, as is partly envisaged in the minds of nature conservationists, met with expressions of more-or-less total rejection. The "wilderness" should show evidence of a clear intention, such as incorporation within a structural or nature-conservation concept. Communicating this is crucial for (increasing) acceptance. Otherwise, these areas become associated with constraints or prohibitions on use. Both would be assigned to the local authority, which is seen as having responsibility, because even "urban wilderness" is seen as local-authority nature (see above).

The most important conclusion from the findings was the following: Terms such as "urban wilderness" or "wilderness in the city" should be avoided both from an urban planning and a nature-conservation perspective. If they are used, they should be linked to clear definitions and concrete objectives. It is a false illusion to believe that one can fence off nature in the city and leave it to become gradually wild or simply enable it to become overgrown as an urban development policy. The nice idea of leaving nature to itself ignores the problem of shrinking cities. Overgrown areas in an environment of emptiness, decay and deconstruction or demolition are not seen as increasing attractiveness. In fact they intensify the already existing impression of dilapidation. Familiar elements from the traditions of parks and gardens should be utilised in the structuring of open spaces in this environment. Also, communication about the new greenery should be linked to motifs that have been handed down and are positive.

A communication strategy for urban nature conservation in general and for fallow nature or "wilderness" in particular could or should therefore be linked to aesthetic motifs and not to ecological arguments. The wilderness concept should not evoke any associations and claims that cannot be satisfied. It must also be made clear that the wilderness concept becomes even more enigmatic in an urban context than it actually is. Fallow nature would therefore be put in semantic competition with other natures where it can only be expected to come off worse.

Acknowledgements

This paper was written as part of the joint project "Urban nature – descriptions of requirements, strategies and action for the management of nature in urban landscapes" which was coordinated by the Department of Urban Landscapes at the UFZ. The author would like to thank Monika Wächter, Anke Jentsch and Axel Philipps for their suggestions, criticisms and comments.

References

Brand K-W, Fischer C, Hofmann M (2003) Lebensstile, Umweltmentalitäten und Umweltverhalten in Ostdeutschland. UFZ-Bericht, Leipzig

Erdmann K-H (2002) Perspektiven des Naturschutzes. Erfordernisse und Möglichkeiten der Weiterentwicklung. In: Erdmann K-H, Schell C (eds) Naturschutz und gesellschaftliches Handeln. Bundesamt für Naturschutz, Bonn, pp 213–241

Hard G, Kruckemeyer F (1993) Die vielen Stadtnaturen – über Naturschutz in der Stadt. In: Koenigs T (ed) Stadt – Parks. Campus, Frankfurt/M. New York, pp 60–69

Kuckartz U (2002) Umweltbewusstsein in Deutschland 2002. Ergebnisse einer repräsentativen Bevölkerungsumfrage. Umweltbundesamt, Berlin

Nymphius J, Trust R (2001) Stadtsafari. Natur entdecken in der Stadt. kbv, Luzern Aarau

Rink D (2002) Naturbilder und Naturvorstellungen sozialer Gruppen. Konzepte, Befunde und Fragestellungen. In: Erdmann K-H, Schell C (eds) Naturschutz und gesellschaftliches Handeln. Bundesamt für Naturschutz, Bonn, pp 23–39

Rink D (2003) Ersatznatur – Wildnis – Wohnstandortfaktor: Soziale Wahrnehmungen und leitbildhafte Vorstellungen von Stadtnatur. UFZ-Diskussionspapier 5, Leipzig

Wächter M (2003) Die Stadt. Umweltbelastendes System oder wertvoller Lebensraum? Zur Geschichte, Theorie und Praxis stadtökologischer Forschung in Deutschland. UFZ-Bericht 9/2003, Leipzig

Nature for People: The Importance of Green Spaces to Communities in the East Midlands of England

Simon Bell

OPENspace Research Centre, Edinburgh College of Art

Introduction

Organisations involved with nature protection or the conservation of biodiversity are generally interested in wildlife and in meeting the requirements of legislation on biodiversity. Recently, organisations such as English Nature, the government agency responsible for biodiversity protection in England, have been given responsibility to obtain data on how the environment contributes to people's social well-being and quality of life.

In urban landscapes, particularly those that have been disturbed by large-scale industrial processes, natural areas defined in strict ecological terms hardly exist. In Britain there is a significant amount of land that has been disturbed to a greater or lesser degree over the last 200 or more years as a result of many industrial processes. Much of this land has been recycled for other uses including agriculture, housing, industry, open space, parks and woodlands. In some places natural regeneration into early successional woodland has taken place but usually quite a heavy series of interventions such as reshaping of the land, drainage, the addition of soil or soil forming materials and planting of trees and shrubs have been the preferred methods of restoration to woodland (Moffat 1997). In a number of areas in Britain industrial or extractive land uses have been located in rural areas where the restoration has been able to create links with other woodlands existing in the landscape.

This chapter explores the importance of nature to the people living in the East Midlands of England, in terms of its social as opposed to biodiversity or economic value, as experienced through a range of publicly accessible green spaces. The research was undertaken for English Nature by the OPENspace Research Centre based at Edinburgh College of Art/Heriot Watt University. The research project took place over the spring and sum-

Kowarik I, Körner S (eds) Wild Urban Woodlands.

mer of 2003 using a combination of qualitative and quantitative research methodologies.

The aim of the project as specified by English Nature, was "to specify the contribution that 'nature' in green spaces makes to people's social well-being by examining the use people make of, and the feelings that they have towards, a selected number of artificial and more natural green space sites distributed throughout the East Midlands". As this was a regional study, the sites were selected to fall more or less equally in each of the region's counties.

The East Midlands is a region of mixed land use and landscape. It comprises the counties of Nottinghamshire, Derbyshire, Leicestershire, Lincolnshire and Northamptonshire. Lincolnshire is largely an agricultural county with a coastline and Derbyshire includes the Peak District National Park. All the counties apart from Lincolnshire have areas of significant extractive industry, especially deep coal mining and open cast iron ore working. These have left large areas of disturbed land which have often been restored to woodland. Several large cities or towns can be found in the region, with significant proportions of ethnic minorities among their populations. The green spaces available to these populations range from waste land used informally; formal city parks, many dating from the Victorian era; country parks, often established on former industrial land; remnant woodlands now lying within the boundaries of urban areas; nature reserves of woodland, wetland, heathland and coastland; and open upland moorland (in the Peak District).

Methodology

The approach used in this research can be described as "user-led" and was based on Personal Construct Theory (Kelly 1955) and the use of Facet Theory (Canter 1977; Shye 1978; Shye et al.1994; Borg and Shye 1995) in the development of the questionnaire. This approach typically starts by exploring issues with members of the public, especially those representing target groups, for example those of concern to policy makers, such as elderly or disabled people, members of ethnic or socially disadvantaged groups etc. The issues raised by these groups are then used to develop questions applied to a wider population. Contrasting views can be obtained by discussing the same issues with professionals involved in the field of research.

For this project this methodological model was applied through the exploration of key issues with members of the public carried out through a number of focus groups located in different parts of the East Midlands area. The results of the focus group research informed the development of a questionnaire used for data gathering from members of the public visiting a number of different "green" areas widely distributed around the East Midlands. A scoping meeting (another form of focus group) was also held with people concerned with nature conservation, including the management of sites such as nature reserves, woodland and parks.

The central objectives of the project as provided in the brief were used as starting points from which to develop the focus group discussions and to understand key issues that were raised. Once identified, usually as the result of the frequency with which people raised the same issue in different groups, the key issues concerning the use people make of, and the feelings they have towards, green spaces were classified into three categories as follows:

- The physical aspects of green spaces
- The activities that people engage in related to green spaces
- The perceptions that people have about green spaces

Table 1 shows these categories and the issues raised in the focus groups which formed the basis of the questionnaire. The questionnaire was constructed from the issues listed in Table 1 using the Facet approach. This uses a "mapping sentence" to construct statements to which interviewees are requested to state levels of agreement or disagreement, along a numerical scale. The scores obtained from this scale then permit statistical analysis. The construction of a mapping sentence is shown in Fig. 1.

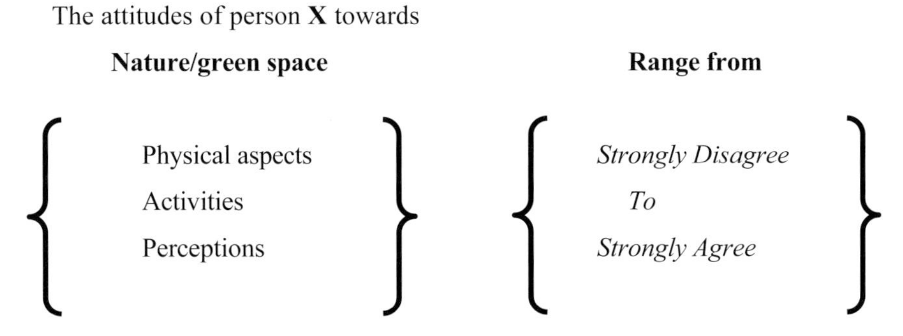

Fig. 1. Construction of the mapping sentence using Facet Theory. Person **X** is defined using the background information

Table 1. Main issues that people raised about the use they make and the feelings they have towards green spaces, from the focus groups

Physical aspects of the green space	Activities that occur in green spaces	The Perceptions that people hold about green spaces
Information about nature is present	Education	Spiritual
Tidiness	Walking alone	Magical
Urban/Rural	Relaxation	Boring
Advertised	Viewing wildlife	Peaceful
Accessible	Exercising	Feel "free"
Man made	Stress relief	Feel "vulnerable"
Proximity to home	Meeting people	Feel "energetic"
Signage to direct to site	Community events	Close to nature
Maintenance	Conservation work	Commercial
Wardens present	Being reminded of childhood places	Owned by community
Well known to individual		Important
		Adventure
		Vandalism
		Comfortable
		Relevant to lifestyle

Results

Scoping meeting

A number of themes emerged from the meeting. Many people shared similar concerns and opinions regarding the subject of the study. Most expressed the view that there is not a single definition of nature, as it depends on a person's educational, ethnic and cultural background. However, they all agreed that the definition of "nature" should not be limited to the physical environment, since it includes "anything that is living", and that the term is wider than "wilderness". They also stressed that nature should not be always associated with the countryside as the former is wider and more embracing than the latter.

Another recurring theme was the social benefit of nature. Attendees listed a wide range of social benefits such as flood management, water quality, recreation, health and wellbeing, arguing that nature can break down barriers by being available to everyone. They realised, though, that there can be an elitist quality in gaining access to nature, as access to some areas has been restricted to long-standing, close-knit groups. Until very recently, many nature reserves were seen as 'out of bounds' and this is still sometimes the case. Fortunately, the situation is improving and wider sections of the community will now visit nature reserves regardless of this perceived elitism or exclusivity. In the East Midlands there are large areas of intensively managed, privately owned farmland with little public access, which leads to an attitude that such places are sterile. As a result, nature has less value in people's minds in the East Midlands.

In conclusion, everyone agreed that nature contributes to the quality of life by making people feel good, giving them a sense of place and an experience that cannot be derived elsewhere. Nature provides a vitally important sense of freedom from the stresses of modern life caused by offices, deadlines, computers, traffic congestion, noise and consumerism.

Focus groups

The main purpose of the focus group research was to gain a qualitative insight into the ways in which people value nature in the study area, and to inform the questionnaire survey designed to cover a wider geographical area. The location of each group and potential target populations (namely the general public but, in particular, to include people with disabilities, minority ethnic groups, women, the elderly and young people) were agreed by the client and the steering group before the inception of the project. The groups took place in six different locations across the East Midlands: Nottingham, Leicester, Mansfield, Corby, Matlock and Spilsby. The focus group discussions were recorded and analysis of these discussions was undertaken by comparing the opinions of the different groups and the frequency with which certain issues were raised across each of the groups. The qualitative nature of the results was reinforced by the inclusion of quotations from some of the group members.

Key points from the discussion of "what is nature?" and "what is green space?" are summarized below:

- The terms "nature" and "green space" are very hard to define.
- Definitions are influenced by cultural perceptions of the natural environment.
- Nature cannot be considered in isolation from the world of human activity.
- Green space can be land over which residents feel they have little or no control.
- Green space can be a small pocket of land in an urban area that is badly maintained and unsafe to use.
- Green spaces can also be very precious.

Key points from the discussion on "what is social benefit?" are as follows:

- The key forms of anti-social behaviour are fly-tipping, litter, vandalism, dogs (mess and running loose) and intimidation from large groups of young people.
- Anti-social behaviour can prevent the implementation of green initiatives.
- Management must be visible whilst at the same time being sensitive to the location.
- There is currently an imbalance between preservation and access to sites of special interest.
- Children are not encouraged to explore and take an interest in nature.
- Parental attitudes towards, and ability to undertake, nature education have changed significantly over the last 50 years.
- The educational system must take responsibility for nature education.
- There is a lack of effective interpretation.
- Green initiatives instill a sense of ownership and encourage responsible behaviour.

Key points from the discussion about the importance of having green spaces nearby are as follows:

- There are many social, mental and physical benefits that can be derived from access to nature and green spaces.
- All the participants felt that access to nature was important, although in some cases the knowledge of nearby nature and green spaces was enough to instill a sense of wellbeing.
- Members of minority ethnic groups are rarely approached to take part in green initiatives and are unsure of where to obtain information.
- Sign posting and information given at sites is often inadequate and not very informative.
- All attempts to provide inclusive access should be sensitive to the location.

The following quotes give a flavour of the way people expressed their feelings about nature:

"Isn't it mean to control somebody through their adolescence to the rooms of the house and the immediate area of the street outside? It's inhumane, it's like imprisonment and I think green open space is a place that they should be able to get into and use. [...] You can't manufacture it; it needs to be random space that they find themselves just as we did when we were young ... a place to light bonfires {laughs}" (Elderly Male, Mansfield, 13/03/2003).

"I think you work better if you've got some green space surrounding you, a few trees and that. Probably at lunch time, it doesn't have to be a big area and you can go and sit on a bench and have your sandwiches rather than be forced to stay inside, and I think it's very beneficial. I think you probably work better after you've had your lunch or your tea break, it doesn't have to be a large area" (Male, Mansfield, 13/02/2003).

"The belief is that if you plant a tree or a sunflower then you are bound to look after it because it is your baby... Sometimes you get minibuses to take people to the countryside, I mean what do they expect people to do; you can't take them to the countryside and just leave them. You have to put it in context. For people to look after the environment it has to be in context with where they are coming from" (Female, Leicester, 22/03/2003).

Questionnaire survey

The questionnaire data was collected at 16 different sites around the east Midlands. The selection was made from a candidate list provided by members of the East Midlands Regional Biodiversity Action Forum and selected to represent a geographical spread and a spectrum from the "wild" to the "urban". These included sites in the Peak District national park, nature reserves, country parks, woodlands, town/city parks and small-scale local green spaces.

Fig. 2. Chaddesden Wood, Derby is an old broadleaved woodland heavily used by local people

Fig. 3. Bourne Woods lie on the outskirts of a small country town and are considered as the local community woodland

Fig. 4. Victoria Park in Leicester is a typical Victorian formal urban park

Fig. 5. Brocks Hill country park is a newly developed area on the edge of Leicester, using a combination of restored derelict land and old farmland

Over 460 interviews were carried out. The interviewers asked people to respond as they left each site following a visit. A target of 30 interviews was set as a minimum. A range of people was wanted, so the interviewers tried to balance the age, gender and other attributes of the sample as they selected people. The questions were, as described above, devised as a set of statements which people were asked to agree or disagree with using a seven-point scale. The data was analysed using the analysis package SPSS. The analysis examined the make-up of the sample in comparison to the poulation as a whole, to see who visits these sites and who does not, then the main attitudinal questions were examined in terms of different demographic variables. Those that proved to be significant as univariate statistics (with a Kruskall Wallis P value of less than 0.05) were examined in more depth. A factor analysis was also carried out across all the attitudinal data to see what groupings of themes emerged. Ten factors were identified, such as "lifestyle", "relax/nature", "welcome" or "childhood/community".

The green spaces used in the survey were chosen to be very different in character. Figures 2–5 show some of the sites.

Results and conclusions

What has been discovered about the social value of nature to the people of the East Midlands of England?

Who is visiting and what do they do?

Many people visit all type of sites, regardless of age or sex. However, there are disproportionately low numbers of people from black and ethnic minorities and people with disabilities. While many people visit on their own, couples and families make up the majority of visitors, the latter especially at the country parks and other sites with special facilities and animals or birds.

Women visitors are under-represented in comparison with the general population, and children formed a smaller proportion than might have been expected given the times of survey. Comparatively low numbers of unemployed people visit; those in employment are mainly in lower supervisory and technical occupations or lower managerial and professional occupations. Many retired people also visit green spaces. The findings about women seem to confirm previous studies that found that women tend to be significantly less frequent visitors than men to woodland or countryside sites (Burgess 1995; Ward Thompson et al. 2002). It may reflect the concerns expressed by women in the focus groups over safety, and women's responses in the attitudinal section of the questionnaire, where feelings of vulnerability were also rated strongly.

Teenaged children are also infrequent visitors compared with younger children. One of the possible causes of this is that what urban teenagers frequently consider "outdoor" places to visit are in fact indoor spaces such as arcades and malls (Travlou 2003). It may be a particular phenomenon of this age group. Læssøe and Iversen (2003), in an in-depth qualitative study of the importance of nature in every-day life, found that youth generates a discontinuity with the nature relationships of childhood because a lot of energy is put into social relations during this phase. The findings also bear out other research into the relationships teenagers and children have with outdoor places such as woodlands (Bell et al. 2003, 2004).

There were few people from black and ethnic minorities visiting any of the green spaces. This seems to follows a common pattern in the UK, as there is a range of evidence from the literature that black and minority ethnic communities in Britain do not participate in visiting the countryside and other natural open spaces, and related activities, proportionate to their numbers in society.

Furthermore, fears of racial and/or sexual attack, of being alone in an unfamiliar environment and worries regarding dangerous flora and fauna, all seem to contribute to a sense of unease in countryside and other natural open spaces (Chesters 1997; Groundwork Blackburn and Manchester Metropolitan University 1999; Slee et al. 2001).

The main reasons people visit green spaces are to walk the dog, to gain exercise, and for the pleasure of being in a park or close to nature. Dog walking is most popular at local sites and in woodlands, also at country parks, but less frequent at nature reserves. Reducing stress and relaxing are significant reasons for visiting green spaces and represent one of the main social values. The importance of dog walking in relation to green spaces has been corroborated by other studies (Ward Thompson et al. 2002; Countryside Agency survey 2003), and cannot be underestimated. A study by Bauman et al. (2001) found that 41 per cent of dog owners walk, on average, 18 minutes per week longer than people without dogs and that if all dog owners regularly walked their dogs, the resulting boost in physical fitness across the community would save Australia's health care system about $175 million every year. In this study, focus groups identified dog fouling as being a key form of anti-social behaviour, so the tensions found elsewhere between dog-owners and other green space users seemed to surface here too (Ward Thompson et al. 2002). One of Tidy Britain Group's surveys found that 80% of people questioned were "greatly concerned" by dog mess, an indication that problems caused by dog fouling are all too common (Tidy Britain Group 1999) and some type of balance has to be achieved. This, however, is not the only problem associated with dogs; a study by Madge (1997) showed that the fear of coming into contact with animals, and in particular dangerous dogs, was much higher for African-Caribbean and Asian groups than white groups.

Many respondents were members of conservation organisations but do not necessarily take an active part in conservation activities.

People think of nature in quite a broad way. They find the term “green space” a difficult term. Nature includes physical characteristics, wildlife and also perceptions and emotions, especially peacefulness and other terms associated with the calming or de-stressing value of nature. Professionals have contrasting views of the distinction between “nature” and “countryside”, for example, and they use the term “green space” more widely than the public understanding of the term.

Getting away from stress was associated with relaxation and nature – seeing it, being in natural places and learning about it. This suggests that there is a role for natural areas for stress reduction, reflected in other studies where it has been shown that leisure activities in natural settings or exposure to natural features have important stress reduction or restoration effects (Ulrich 1981, 1993; Sheets and Manzer 1991; Kaplan 1995; Parsons et al. 1998; Ulrich et al. 1991).

When talking about “social values” people tended to focus on “anti-social uses”. There is a lot of evidence that sites need to be well managed (but not over managed), welcoming, provide information and have a natural appearance if people are to obtain the best value from them.

Sites close to home are preferred, especially by those who used to visit frequently when children. This importance of accessibility to places close to home compared with the site character is reflected in other research (Ward Thompson at al. 2002).

There are significant associations between the type and degree of use of green spaces by people now and how frequently they visited such sites when children. This suggests that if children are not being allowed or encouraged to visit natural areas or other parks by themselves, they are less likely to develop a habit that will continue into adulthood. Those who had visited a lot as children were more likely to find magical and other positive qualities in nature, and to develop a closer relationship with it as part of their lifestyle, than those who did not.

Accessibility and welcome were rated highly and this seems to go with a sense of community ownership of green space, when there is a sense that it belongs to the community as much as to the formal or legal owners.

The sense of feeling uncomfortable or vulnerable was not very widespread overall, although it was most significant among the female and older respondents. This sense of vulnerability among women reflects the findings of other research (Burgess 1995; Ward Thompson et al. 2002). An international example of ways of dealing with this issue is the city of Montreal's Women's Safety Audit, which considers that it is vital not only to take into account the specifics of sexes but also the particulars of groups (elderly and disabled people, ethnic and sexual minorities) as well as involving men in their role of father, partner, son or potential victim (Michaud 1993).

The sites that attracted most positive responses to perceptions were the nature reserves, woodlands and urban parks. Local areas were important for some activities but country parks tended to score less highly. Responses in relation to nature reserves were very positive compared with most other sites. This is partly the value of their being good for children to learn about nature, but other values, such as being associated with spiritual qualities, getting free from stress and feeling energetic are also positively associated with nature reserves. Woodlands share many of these attributes. Wild areas and country parks have the most associations with being bored but also have some positive values.

Lessons for providers

The research has flagged up a number of useful areas for consideration by providers of nature and outdoor recreation:

How do the findings of this research affect the implementation of strategic environmental assessments, part of a recent EU directive due to be implemented soon?

The importance of different kinds of green space and of easy and welcoming access for all, including children, disabled people and people from ethnic minorities, needs to be taken into account in strategies for the regeneration of derelict areas, alongside other social and environmental needs.

Urban parks were highly rated in this study. What are the implications for their funding, regeneration and management?

The implications raised by the findings for regional environmental strategies need to be considered.

Country parks emerged less favourably from the research than some other areas and there are implications for their future. Are there ways to enhance their social value?

Lessons for managers

There are many pointers to things that managers can do to encourage more people to visit green spaces and to ensure that, once there, the visit is a good one.

More and better information is needed, to tell people where they can go, what they can do and how they can get there, orientated towards different groups, such as black and minority ethnic groups, disabled people, older people, socially disadvantaged people etc. This may need to be in different languages, presented in different ways and distributed differently in order to meet the needs of those not reached at present.

Information at sites is also important, possibly presented in new ways and aimed at different groups in what is clearly a fragmented, not a homogeneous population.

More activities and means of engaging children in green spaces should be considered, so that they develop a habit of visiting them. It is important, nonetheless, to understand why teenagers may not want to visit such sites. Working with parents and police/rangers etc. to develop a safer environment so that children are allowed to go out by themselves would be very helpful.

Further development of educational programmes for children is necessary. This was seen by many people as vital yet also seemed not to be widely enough available. Using green areas near schools, which are easier to visit and not necessarily special parks, should be considered.

References

Bauman A, Russell S, Furber S, Dobson A (2001) The epidemiology of dog walking: an unmet need for human and canine health. The Medical Journal of Australia, 175:632–636

Bell S, Ward Thompson C, Travlou P (2003) Contested views of freedom and control: children, teenagers and urban fringe woodlands in central Scotland. Urban Forestry and Urban greening 2:87–100

Bell S, Morris N, Findlay C, Travlou P, Gooch D, Gregory G, Ward Thompson C (2004) Nature for people: the importance of green spaces to East Midlands Communities. English Nature Research report 567, English Nature, Peterborough

Borg I, Shye S (1995) Facet Theory; Form & Content. Sage Publications, London

Burgess J (1995) Growing in Confidence: a study of perceptions of risk in urban fringe woodlands', Technical report CCP 457, Countryside Commission (former),Cheltenham,http://www.nottingham.ac.uk/sbe/planbiblios/bibs/country/09.html, last visited 20/10/03

Canter D (1977) The Psychology of Place. Architectural Press, London

Chesters A (1997) 'Who's been left out'. In: Blamey E (ed) Making Access for All a Reality. Proceedings, Countryside Recreation Network, Cardiff University: 32–36

Countryside Agency, English Heritage and Sports England (2003) The use of Parks, www.sportengland.org/press-releases/The-Use-of-Parks.pdf, last visited 20/10/03

Groundwork Blackburn and Manchester Metropolitan University (1999) Involving Black and Minority Ethnic Communities. Groundwork Blackburn and Manchester Metropolitan University

Kaplan S (1995) The Restorative Benefits of Nature: Toward an Integrative Framework. Journal of Environmental Psychology 15:169–182

Kelly G (1955) Principles of Personal Construct Psychology. Norton, New York

Læssøe J, Iversen T (2003) Naturen i et hverdagslivs-perspektiv: En kvalitativ interview-undersøgelse af forskellige danskeres forhold til naturen. Faglig rapport fra DMU, nr. 437

Madge C (1997) Public parks and the geography of fear. Tijdschrift voor Economische en Sociale Geografie 88 (3):237–250

Michaud A (1993) Agir pour une ville sans peur. Guide d'enquête sur la sécurité des femmes en ville, City of Montreal, Department of Leisure, Parks and Community Development. From: VIOLENCE AGAINST WOMEN IN URBAN AREAS, An analysis of the problem from a gender perspective

Soraya Smaoun United Nations Centre for Human Settlements, Urban Management Programme Safer Cities Programme Women and Habitat Programme, Nairobi, October 1999, last visited 16/10/03
Moffat AJ (1997) Recycling land for forestry. Forestry Commission Technical paper 22. Forestry commission, Edinburgh
Parsons R, Tassinary LG, Ulrich RS, Hebl MR, Grossman-Alexander M (1998) The View from the Road: Implications for Stress Recovery and Immunization. Journal of Environmental Psychology 18:113–140
Slee B, Curry N, Joseph D (2001) Social Exclusion in the Countryside Leisure in the United Kingdom: The Role of the Countryside in Addressing Social Inclusion. A Report for The Countryside Recreation Network. The Countryside Recreation Network
Sheets VL, Manzer CD (1991) Affect, Cognition, and Urban Vegetation: Some Effects of Adding Trees Along City Streets. Environment and Behavior 23(3):285–305
Shye S (ed) (1978) Theory Construction and Data Analysis in the Behavoiral Sciences Jossey-Bass Publishers, San Francisco
Shye S, Elizur D, Hoffman M, (1994) Introduction to Facet Theory; Content Design and Intrinsic Data Analysis in Behavioral Research Sage Publications, London
Tidy Britain Group (1999) Dog fouling and the law. Published by DEFRA. Page published 1 June 1999, last modified 4 April, 2003, visited 20/10/03, http://www.defra.gov.uk/environment/localenv/dogs/foul.htm
Travlou P (2003) People and Public open Space in Edinburgh City Centre: Aspects of Exclusion. Presented at the 38th International Making Cities Liveable Conference, Carmel, California 19-23 October
Ulrich RS (1981) 'Natural versus urban scenes: some psycho-physiological effects'. Environment and Behaviour 13:523–556
Ulrich RS (1993) Biophilia, Biophobia, and Natural Landscapes. In: Kellert S, Wilson EO (eds) The Biophilia Hypothesis. Shearwater/Island Press, Washington DC, pp 74–137
Ulrich RS, Simons RF, Losito BD, Fiorito E, Miles MA, Zelson M (1991) Stress Recovery During Exposure to Natural and Urban Environments. Journal of Environmental Psychology 11:201–230. From: The Calming Effect of Green: Roadside Landscape and Driver Stress, fact sheet no 8, University of Washington, College of Forest Resources, Human Dimensions of the Urban Forest www.cfr.washinton.edu/research.envmind/Roadside/Rsd-Stress-FS8.pdf, last visited 20/10/03
Ward Thompson C, Aspinall P, Bell S, Findlay C (2002) Local open space and social inclusion: case studies of use and abuse of woodlands in Central Scotland. Edinburgh: OPENspace research centre report to Forestry Commission.

Living in the Urban Wildwoods: A Case Study of Birchwood, Warrington New Town, UK

Anna Jorgensen, James Hitchmough, Nigel Dunnett

Department of Landscape, Sheffield University

Introduction

Relevance and context of the study

This research investigated urban dwellers' experience of living in a woodland setting on a post-industrial site using Birchwood (part of Warrington New Town, UK) as a case study. It is relevant to any agency concerned with landscape planning in the peri-urban area in localities where woodland is, or could be, the dominant landscape element.

In the UK large scale urban landscape planning had its heyday as part of the New Towns Movement from 1945 to the early 1980's, when the then Conservative government moved housing provision away from the state into the private sector. Currently the British government is once again dealing with housing provision at a more strategic level by looking to promote house-building in four regional locations or "growth areas" (Office of the Deputy Prime Minister 2003). Yet if woodland is to be considered as one of the principle means of structuring and integrating these new settlements it is essential to find out whether this would meet with public approval. In particular, if people disliked or were afraid of naturalistic woodland in an urban setting, it seems unlikely that it would have the psychological, social and health benefits that are beginning to be associated with other forms of urban nature and green space (Rohde and Kendle 1994).

The aim of the study was therefore to evaluate residents' reaction to Birchwood's woodland landscape with a view to contributing to the debate about the appropriate form and qualities of urban green structure and green spaces in the context of current plans for urban development and expansion.

Kowarik I, Körner S (eds) Wild Urban Woodlands.

The urban woodlands at Birchwood

Birchwood was constructed on the site of the former Royal Ordnance Factory at Risley, built to supply armaments during the Second World War. Constructed on a largely "green field" site, the factory employed 30,000 people and was operational for just seven years. After the war the 740-hectare site lay disused and derelict for twenty-five years. To the south lay a belt of woodland running alongside the Liverpool to Manchester railway line, and to the south west lay an area of woodland and raised bog known as "Risley Moss", the remnant of a much larger area of bog that had originally covered the site itself and surrounding area. During this 25 year period natural succession took over and the site was colonized by *Quercus, Fraxinus, Betula* and *Salix* species, amongst others. The factory was demolished and the settlement of Birchwood was constructed during the 1970's and early 1980's. It consisted of three housing areas, "Oakwood", Gorse Covert, and Locking Stumps, as well as an employment area and shopping centre. By 1991 Birchwood was home to 12,500 inhabitants housed in a mixture of privately owned and rented dwellings (64/35%). The 1991 census returns indicated that 38% of the "heads of household" were employed in "managerial and technical" and 19% in "skilled manual" occupations.

The development at Birchwood was unusual in the sense that it was landscape driven, and the concept for the landscape plan was to use woodland as the main structuring element. Rather than being confined to isolated patches, the woodland was a continuous entity surrounding and flowing through the whole development, dividing up parks and green spaces, running alongside roads and footpaths, and creating a matrix into the cells of which the built development was placed.

Early on, a decision was made to retain Risley Moss and its surrounding woodland, as well as the belt of woodland alongside the Liverpool to Manchester railway line, and other isolated vegetation outcrops. The different vegetation types found in these woodlands were used to develop four basic vegetation mixes, adapted to variations in local conditions such as differences in substrate and the availability of light, which were used to create the woodland structure on site. Thus the newly planted woodlands at Birchwood were in some respects "grown on" to the existing woodlands, and retained much of their character.

There were many aspects of the methods at Birchwood that can be described as "ecological", and the whole approach was widely imitated in Britain and became known as "the ecological style" or "the ecological approach". The ecological woodland approach at Birchwood was inspired partly by contemporary ideas from Europe, particularly from Sweden and the Netherlands, but was in many ways the largest and most uncompromising example of the application of these ideas.

Location on the urban-rural gradient and type of nature

The Birchwood woodlands can be defined as "woodlands in the urban fringe". In terms of the typology developed by Kowarik (2005) they are a mix of "Nature of the third kind" (planted tree stands in green spaces and other forms of functional greening) and "Nature of the fourth kind" (woodlands naturally established on specific urban-industrial sites).

In this paper the woodland at Birchwood is frequently described as "naturalistic". This expression was often used by Birchwood's planners and designers (Tregay and Gustavsson 1983). Judging by the manner and context in which these expressions were used and by the landscape itself, Birchwood's new landscape was "nature-like" or "naturalistic" because it was informal, with an organic structure, and made use of ecotones such as woodland edge as transitions between different plant communities and vegetation types; and because it resembled the spontaneously occurring existing woodland and open scrub on and around the site. Thus, whenever the word "naturalistic" is used to describe woodland or vegetation in Birchwood in this paper, these are the characteristics that are implied.

Literature review

One of the difficulties in evaluating a more naturalistic approach to urban tree plantings in the UK is that such an approach was rarely used as part of planned or designed urban landscapes before Birchwood. Tartaglia-Kershaw (1980) carried out a study of the Gleadless area of Sheffield, in the UK, a housing area planned around existing mature woodland. Although the woodland in Gleadless was generally within 500 metres of the housing, and often considerably closer, it was not closely integrated with the housing, as in Birchwood. Seventy two per cent of the sample in the study said that the woods were important to them. An overwhelming 90% liked living on the estate, and 94% said that they liked the way the area had been planned. However, Tartaglia-Kershaw (1980) concluded that the overall findings did not support the approach used at Birchwood:

> "The residents do not want woodland to the door as many figures in the "Nature in Cities" movement suggest, and which is happening in New Towns based on woodland structure planning (sic)."

Burgess et al. (1988) found that traditionally managed urban green spaces characterized by isolated trees and mown grass were not valued as much as natural or semi-natural urban landscapes characterized by woodland, multiple layers of vegetation and an un-mown grass/herb layer. However, they also found that many people had ambivalent feelings about the landscapes they most valued: these landscapes were also the ones that aroused the most fear. In this respect they went further than other commentators such as Kaplan and Kaplan (1989), who had suggested that "mystery" was an important component of landscape preference; but who did not go on to explore the adverse responses that many people experience when faced with wild urban nature in any detail.

Although Valentine's (1997) research was carried out in a rural context it exposed the contradictory notions that parents hold regarding their children's safety in green settings generally: on one hand such settings are seen as good places to bring up children, yet on the other, such surroundings are also perceived to be fraught with danger.

Burgess' findings about the value that people place on natural or semi-natural urban landscapes were confirmed and explored in more detail by Bussey (1996), who found that woods were ranked above parks, and second only to open countryside, as the preferred landscape for informal recreation.

Despite the innovative work done by researchers such as Burgess and Bussey, the idea that "woodland structure planting" is regarded as unsafe by members of the general public, and is therefore unsuitable for use in urban situations has persisted (Thompson 2000). There is clearly a danger that, in seeking to reassure the general public by the removal of dense woody vegetation, we are also destroying the landscapes that people most value, despite their understandable fears. However, it may also be the case that "the Ecological Approach" was too wholesale, in that naturalistic vegetation was used too indiscriminately and too close to people's homes, as predicted by Tartaglia-Kershaw (1980). There may well be an appropriate gradient of planting styles, ranging from formal and manicured to wild and nature-like.

In his study of the use of woodland in conjunction with housing, Dowse (1987) recommended an interface between the dwellings and the woodland containing a parkland zone at least 500 metres wide comprising "clumps and specimen trees set in drifts of shrubs at a distance from housing". Manning (1982) has also advocated an appropriate gradient between "intensive" and "extensive" landscapes, though unlike Dowse (1987), he does not seek to prescribe particular types of vegetation in particular locations or lay down strict guidelines. Arguably what is needed is an element of choice, as proposed by Burgess et al. (1988). People may welcome more naturalistic treatments provided they can choose when to interact with them.

The key issues that emerge are the conflict between human appreciation for urban nature and the fears that urban green space (and especially naturalistic or wild-looking landscapes) can engender in urban dwellers, for themselves and for their children, as well as the implications for the planning and design of new urban areas.

Research issues

The study sought to address these issues of aesthetic appreciation for the woodland, perception of personal safety, and the perceived suitability of Birchwood as a place to bring up children, as well as exploring these issues in more detail by examining the cultural meanings that the woodland held for Birchwood's inhabitants. In this context "cultural meanings" means the collective understandings held by a set of people with common characteristics (e.g. place of residence, gender). These may be developed or altered by individual reflection or experience (Bourassa 1991), and may differ widely between societal groups with different interests or characteristics.

Methodology

Overview of research design

Birchwood was selected as a case study because, in Britain at least, it is the largest and most radical example of the ecological woodland approach. The principal research instruments were a postal questionnaire, and a series of semi-structured interviews with a sub-sample of the questionnaire respondents. The sampling strategy was to obtain a random stratified sample. The strata were nine "housing character areas" ("HCA's") representing a typology of different residential areas in Birchwood, with differing vegetation and housing densities. There was also a "control" sample drawn from three HCA's outside Birchwood (the method of selection of these HCA's is outlined below).

Identifying the HCA's

Birchwood is not a homogeneous entity. The residential part alone encompasses a variety of housing tenures and types, in diverse layouts. The amount of vegetation varies considerably from one area to another and in some parts, the woodland is very closely integrated with the housing (Fig. 1). It seemed possible that these differences would have an impact on public perception of the research issues. For these reasons it was decided to develop a typology of the residential areas in Birchwood that would reflect these differences. This typology was then used to select the areas from which the questionnaire sample and the interviewees would be drawn, using the vegetation density and the housing density of the areas as the criteria for selection. These two criteria were chosen because they were considered to be the main overall differentiating factors between the various areas. It was considered that housing density was the best indicator not only of the spacing and layout of dwellings, but also of their size and type.

Fig. 1. In some parts of Birchwood the woodland is closely integrated with the housing

Vegetation density and housing density

The HCA's were identified by means of an urban landscape character assessment of Birchwood. The vegetation density of each HCA was measured using a technique derived from small-scale vegetation mapping and measuring techniques, based on the work of the ecologist Braun-Blanquet (Kent and Coker 1992). The area of each HCA in hectares was divided by the number of dwellings it contained to give a figure for housing density measured in dwellings per hectare.

Selection of sampling HCA's

The urban landscape character assessment yielded a total of 33 HCA's. In order to obtain the widest possible range of vegetation and housing densities within the HCA's to be sampled, three conditions of each of these variables were selected, namely high, medium and low. Putting these conditions together in all possible combinations gave rise to nine pairs of conditions, or cells. The data for vegetation and housing density was then scrutinized to find the nine HCA's (out of the original 33) most closely matching these nine combinations.

Selection of control HCA's

In order to control for differences in perception attributable to the presence or absence of the woodland setting it was decided to compare the perceptions of Birchwood residents with those of residents of areas without this setting, as well as making comparisons between HCA's in Birchwood with different characteristics. It was therefore decided to select three "control HCA's" from the rest of Warrington with low, medium and high housing density, but with little or no woody vegetation.

The postal questionnaire and interviews

The questionnaires were posted to randomly selected residents of the 12 HCA's (nine from within Birchwood and three "control" HCA's from outside). A total of 1181 were sent out and 336 were returned (response rate 28%). Out of these, 266 were from respondents in Birchwood, and 70 from the control sample outside. It was originally intended that there would be at least 30 responses from each HCA. Despite two separate mailings this target could not be met in two thirds of cases due to the small size of some of the HCA's, and the resource constraints of the study. Whilst many of the findings are based on differences between the larger Birchwood and control samples (n=266/70) it is accepted that the small size of some of the HCA sub-samples means that any findings drawn from differences between them must be somewhat tentative.

Subsequently semi-structured interviews were carried out with 39 of the questionnaire respondents drawn more or less evenly from the 12 HCA's.

Methods of statistical analysis

The questionnaire data was analysed using the statistics package, SPSS version 11. The data was coded and transformed into a number of different types of variable namely nominal (binary), nominal (categorical), ordinal and scale. A selection of four different non-parametric statistical tests was used to test for the existence of statistically significant associations or correlations between different combinations (pairs) of variables. The rationale behind the choice of the four tests was that the most powerful and appropriate test available should be used for any given combination of variables. Table 1 indicates which test was used for a particular combination of variables.

Table 1. Statistical tests used on variable combinations

Variables	Nominal (binary)	Nominal (categorical)	Ordinal	Scale
Nominal (binary)	Chi-Square	Chi-Square	Mann-Whitney	Mann-Whitney
Nominal (categorical)	Chi-Square	Chi-Square	Kruskal-Wallis	Kruskal-Wallis
Ordinal	Mann-Whitney	Kruskal-Wallis	Spearman's Correlation	Spearman's Correlation
Scale	Mann-Whitney	Kruskal-Wallis	Spearman's Correlation	Spearman's Correlation

There were essentially three types of variables used in the study namely the independent variables (see following paragraph), the demographic variables (gender, age, occupation and educational attainment) and the dependent, perceptual variables representing the four research themes (more information is given about these in the thematic sections below).

Three independent variables are referred to in this paper and these are "vegetation density", "housing density", and "location in relation to Birchwood". The scores for the vegetation and housing density of the 12 HCA's were simply transformed into the scale variables "vegetation density", and "housing density", where the values consisted of the vegetation scores and dwellings per hectare respectively. The variable "location in relation to Birchwood", was a nominal (binary) variable, where the values 1 and 2 denoted whether the respondent lived in Birchwood, or in one of the three control HCA's outside.

Results and discussion

Selected results are reported and discussed in four separate sections of this paper, corresponding with the research issues or themes, namely aesthetic factors, cultural values and meanings, perception of personal safety and bringing up children and the perception of children's safety.

Aesthetic factors

The postal questionnaire contained a number of questions about the visual appearance of the street where the respondents lived. Respondents were asked:

"Compared to other places you have lived, or other places you know, do you like or dislike the way your street looks?"

The respondents were asked to respond using a five point bi-polar Likert scale, with responses ranging from "Like very much" to "Dislike very much". The data was converted to an ordinal variable with values between 1 and 5, reflecting the five categories on the Likert scale, where 5 was "like very much" and 1 was "dislike very much". Birchwood respondents were slightly more positive about their street than their counterparts from outside: whilst 76% of Birchwood respondents said that they liked the way their street looked, or liked it very much, only 70% of the control respondents did so. However this difference was not statistically significant (Mann-Whitney Z = -0.564; NS).

The respondents were then asked whether they liked or disliked a range of specific aspects of their street including items such as "car parking", "visual appearance of the houses" and "trees and greenery". The data from this question was converted into a series of nominal (binary) variables, where 1 was "like" the aspect in question, and 2 was "dislike" the aspect in question, e.g. 1= like "trees and greenery" and 2= dislike "trees and greenery". Birchwood respondents were significantly more enthusiastic about the "trees and greenery" on their street: 94% said they liked this aspect of their street compared to 85% of the control group (Chi-Square $x^2 = 5.895$; $df = 1$; $p = 0.015$).

Somewhat surprisingly, the vegetation density of the respondents' HCA's had no impact on either their approval for the visual appearance of their street nor their liking for the "trees and greenery"; but there were significant trends for respondents from the higher housing density HCA's to be more disapproving of both, though in the case of "trees and greenery" the result was only marginally significant (Table 2, Fig. 2).

Table 2. Effect of vegetation and housing density on respondents' approval for the visual appearance of their street, and on their tendency to approve or disapprove of "trees and greenery"

Variables	Test	Result
Street/vegetation density	Spearman's correlation	$r_s = -0.030$; $n = 262$; NS
Trees and greenery /vegetation density	Mann-Whitney	$Z = -.167$; NS
Street/housing density	Spearman's correlation	$r_s = -0.340$; $n = 262$; $p < 0.0001$
Trees and greenery /housing density	Mann-Whitney	$Z = -2.030$; $p = 0.042$

These results indicate that the respondents from the higher density social housing both in and outside Birchwood were less satisfied with the way their streets looked. Did they also mean that more of these respondents disliked the "trees and greenery"? The higher level of dissatisfaction with the street found amongst these respondents is unlikely to be connected with housing density itself: there are many examples of high density housing that are highly sought after. One example is the current British fashion for so called "city living" in purpose-built high-density flats, but there are older examples such as the Barbican in the City of London. This trend is more likely to be a reflection of a lower level of satisfaction with the circumstances of daily living connected with factors such as unemployment, poverty, lower levels of educational attainment and ill health.

As Fig. 3 confirms respondents from the high housing density HCA's were more likely to be unemployed, have lower levels of educational attainment, to be single parents or be over 59. Whilst these characteristics are not synonymous with deprivation, where they occur together, as in this example, it seems likely that poverty and deprivation are also present. These four factors are very similar to those used in the Index of Multiple Deprivation (Department of the Environment, Transport and the Regions 2000).

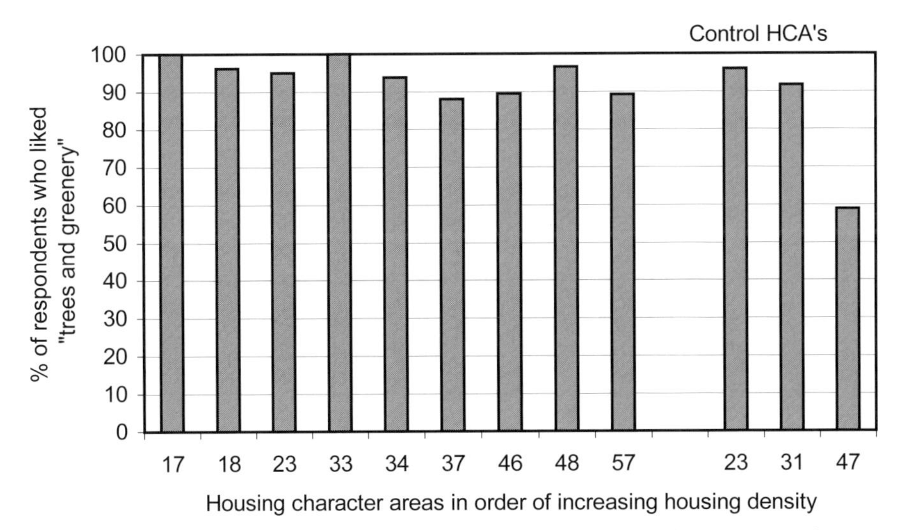

Fig. 2. Effect of housing density on respondents' approval for "trees and greenery"

The interviews confirmed that there were differences in the perceptions of respondents from HCA's with different housing densities; those from higher housing density HCA's felt more dissatisfied with the way in which vegetation was being managed, and less able to take personal control of it. Housing density and housing tenure appeared to mediate respondents' attitudes towards "trees and greenery", through their links with choice of accommodation, size of plot, proximity of peripheral vegetation and ability to manage or control the vegetation. It should be emphasized that there was no real evidence that respondents from the high housing density HCA's disliked "trees and greenery" *per se,* any more than respondents from other HCA's. Rather, it was the proximity of the vegetation and their inability to control it that was seen as problematic. Here the comments of two respondents from high and low density housing respectively are contrasted:

Mrs. Sh: "if they kept it low enough and neat enough there's no problem with them, and if they hadn't planted them so close to, I mean a lot of the bushes that are round the houses are actually planted virtually against the walls, and when they're not maintaining them properly they weigh up the walls, and it cause damp and all sorts to the houses so maybe they should have made a better plan of exactly how far they should have been planted to the house, and to brickwork and how, you know how much maintenance they were going to take in the future because when they were first put in I mean they were only little tiny things weren't they?"

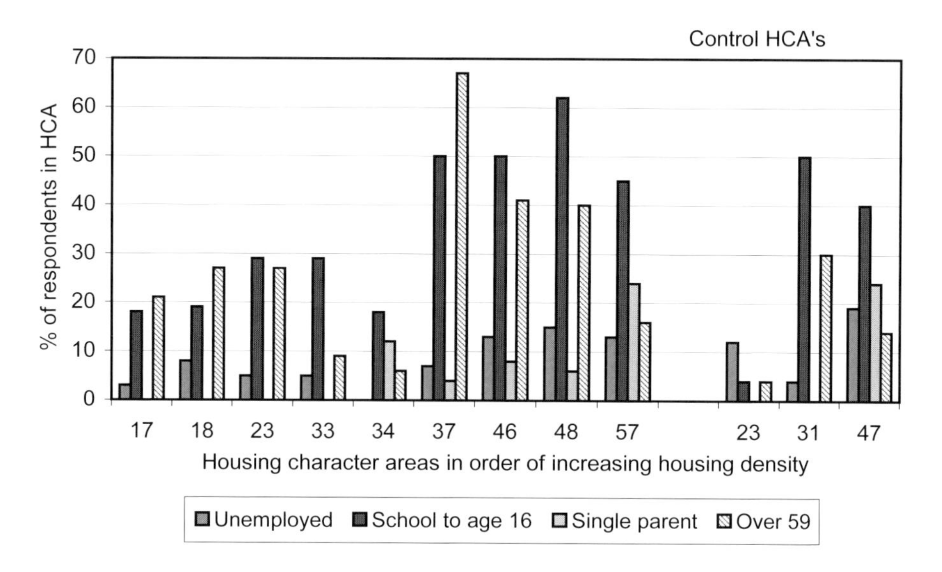

Fig. 3. Indicators of deprivation in housing character areas (source – postal questionnaire)

Mr. Sp: "When I had the path paved, for me the tree was just a bit too big and I was, there was some cotoneaster underneath it as well, and I wanted to make something that was a little bit less obtrusive right in the drive, I wasn't worried about it damaging the roots ultimately affecting the house, I mean that was a reasonable distance away, I used that as an opportunity to get rid of it but I planted lower vegetation since to keep the slightly open aspect, the trees are nice on the edge if you want, we've got a quite small close it's only 13 houses I don't think it can quite cope with too many big trees."

These issues of control and proximity highlighted the need for respondents to *personalise* their own living space: a need that was articulated by a number of respondents during the interviews. Many respondents from low, medium and high housing density HCA's described how they had removed the original planting put in by the builders and contractors at the time their homes had been constructed, usually from their front gardens (Fig. 4).

Cultural values and meanings

In the postal questionnaire the respondents were asked to name up to three places that they particularly liked in their local area, not including their own home and garden. The respondents' first named places were sorted into five categories, namely "green spaces", "outdoor recreational spaces" (e.g. Birchwood Golf Club), "indoor recreational spaces" (e.g. Birchwood

Fig. 4. This Birchwood resident has personalised their front garden by removing the original hedges and soft landscape and replacing them with highly manicured alternatives

Shopping Centre), "footpaths" and "other". "Green spaces" were the places most respondents valued in their local arca (chosen by 63% of all respondents), irrespective of whether they came from Birchwood or from the control areas outside, even where there were competing attractions locally such as shopping centres or the golf club. In Birchwood the most popular green spaces were Risley Moss (chosen by 26% of Birchwood respondents) and Birchwood Forest Park (chosen by 18%). The publicly accessible parts of Risley Moss consist predominantly of woodland (Fig. 5), whereas the Forest Park consists of a series of linked open grassed areas, framed by woodland belts. Both have a strong woodland character.

The high esteem in which these places (and other green spaces in Birchwood) are held confirms the value of woodland as a local recreational resource, as indicated by previous research (Tartaglia-Kershaw 1980; Burgess et al. 1988; Burgess 1995; Bussey 1996).

Fig. 5. Woodland in Risley Moss, Birchwood

The interviews revealed that the respondents from Birchwood valued its green and wooded spaces, and the vegetation and wildlife found in those spaces for a number of reasons, which may be summarised as:

- A feeling or belief that Birchwood was making a precious contribution towards the conservation of nature and wildlife, and that in Birchwood humans can co-exist with nature;
- An awareness of seasonal change;
- The potential to engender experiences of a transcendental nature (e.g. the ability to "lose oneself");
- Rural idyll;
- Relaxation, tranquility and stress relief.

Whilst the naturalistic landscape of Birchwood had its own set of meanings for many respondents these meanings were not necessarily synonymous with Birchwood's identity as a place. This was partly because different respondents had different ideas about Birchwood's physical extent. More importantly, when questioned about Birchwood's identity, most respondents replied by talking about the community, or about social groupings or institutions that represented the community for them.

Fig. 6. The hanging baskets are symbols of a caring community for Birchwood inhabitants

This suggests that the landscape is evaluated according to whether it exhibits positive or negative signs of its inhabitants. Thus, signs of caring human intervention are greatly valued, whereas vandalism and abuse reinforce negative ideas regarding the community, and the landscape.

The interviews suggested that in Birchwood the floral hanging baskets recently installed along the expressways by the Town Council are amongst the most potent symbols of a caring community for Birchwood inhabitants (Fig. 6). This type of embellishment was definitely not part of the original plan for Birchwood, and would undoubtedly have been seen as incongruous by the original designers and planners, who envisaged the expressways running through a wholly naturalistic woodland landscape.

Their strategic location at the roundabouts suggests that the hanging baskets may have another important function. The roundabouts are essentially gateways to Birchwood, marking the transition from the naturalistic woodland belts to the built development, but in the original plan they were not sufficiently differentiated as such. There was certainly an attempt by the designers to vary the rhythm of the tree planting and to locate eye-catching trees species at key locations, but these strategies seem not to have had sufficient impact. It seems that flowers and colour, through their association with caring, have the ability to mark the passage from the wilderness zone of the woodland to the cultivated zone of the built development, and that these kinds of symbols and markers have the potential to perform important transitional functions.

Perception of personal safety

In the postal questionnaire the respondents were asked whether there were any places in their local area where they would feel unsafe alone, during the day time and after dark respectively. The data from these questions was converted in each case into a nominal (binary) variable where 1 was "yes" and 2 was "no" (denoting that there were, or were not, places in the local area where the respondent would feel unsafe alone). The respondents who had answered "yes" to either of these questions were then requested to identify up to three unsafe places in their local area, during the day time and after dark respectively. The respondents' first named places were sorted into eight categories namely: "local facilities" (e.g. local shops), "roads and motorways", "built-up areas" (whole districts that respondents identified as being unsafe, e.g. Oakwood), "large built structures", "pathways, bridges and underpasses", " green spaces" and "other".

The respondents from Birchwood were significantly more likely to feel unsafe in their local area, both during the day time and after dark, compared with the control group from outside (Table 3). Thirty seven per cent of the Birchwood respondents identified unsafe places in the local area during the day time, compared to 23% of the respondents from the control HCA's. After dark, the contrast was more marked: 75% of the Birchwood respondents identified unsafe places, whereas only 54% of the control respondents did. The types of places most commonly identified as unsafe across the whole sample were "pathways, bridges and underpasses" and "green spaces" (Fig. 7).

Female respondents from Birchwood were significantly more likely to identify unsafe places compared to Birchwood's male respondents (Table 4).

Women from Birchwood were significantly more likely to identify unsafe places in their local area, compared to their counterparts from outside (Table 5). It seems probable that Birchwood's woodland setting contributes to this difference in perception.

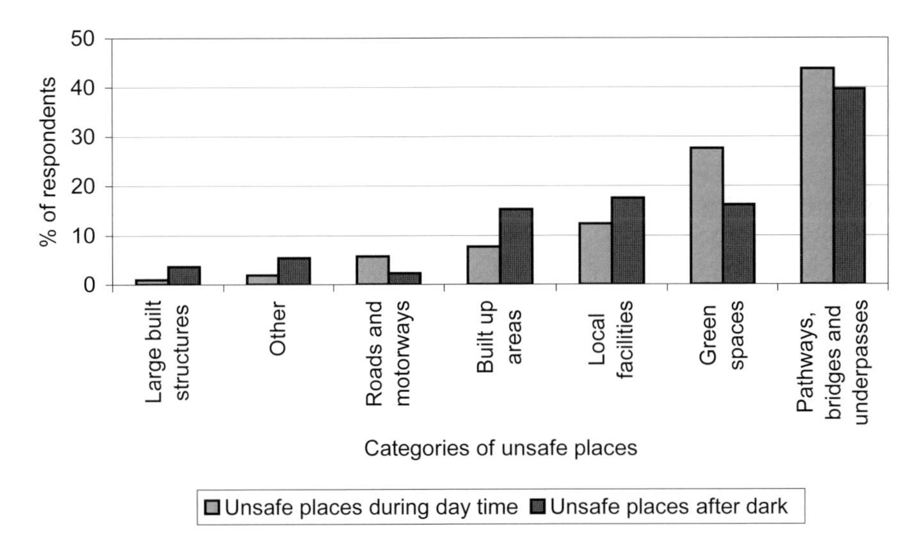

Fig. 7. Respondents' choice of unsafe places in their local area during the day time and after dark

Table 3. Effect of living in or outside Birchwood on respondents' tendency to i-dentify unsafe places in their local area, during the day time and after dark

Day time/after dark	Test	Result
Day time	Chi-square	$x^2 = 4.444$; df = 1; $p = 0.035$
After dark	Chi-square	$x^2 = 11.076$; df = 1; $p = 0.001$

Table 4. Effect of gender on Birchwood respondents' tendency to identify unsafe places in their local area, during the day time and after dark

Day time/after dark	Test	Result
Day time	Chi-square	$x^2 = 24.667$; df = 1; $p < 0.0001$
After dark	Chi-square	$x^2 = 13.674$; df = 1; $p < 0.0001$

Table 5. Effect of location in relation to Birchwood on female respondents' tendency to identify unsafe places in their local area, during the day time and after dark

Day time/after dark	Test	Result
Day time	Chi-square	$x^2 = 10.424$; df = 1; $p = 0.001$
After dark	Chi-square	$x^2 = 10.589$; df = 1; $p = 0.001$

The findings therefore represent a curious paradox. As we have seen, whilst "green spaces" were often thought of as unsafe, they were also the most valued places in Birchwood, confirming that urban dwellers often hold conflicting feeling towards naturalistic or wilderness like places (Burgess et al. 1988). Whilst steps can be taken to make such places feel safer (Burgess 1995; Kaplan et al. 1998) they cannot be made to feel completely safe without removing the qualities that attract people to them in the first place. The existence of these contradictory feelings about nature-like green spaces strongly suggests that we need to re-examine existing models of landscape preference in which preference and safety are seen as co-dependent (e.g. Appleton 1975; Orians and Heerwagen 1992), and this is an interesting area for further research.

The perception of children's' safety, and the perceived suitability of Birchwood as a place to bring up children

In the postal questionnaire the respondents were asked whether they felt that their local area was a good place to bring up children. The data from this question was transformed into a nominal (binary) variable with values 1 and 2, where 1 signified that the respondent's local area was a good place to bring up children and 2 signified that it was not. The respondents were then requested to give reasons for their answer to this question and these reasons were sorted into categories (e.g. "good community" and "local green spaces/green setting").

Respondents from Birchwood were significantly more likely to feel that their local area was a good place to bring up children, compared to the control respondents (Chi-Square $x^2 = 6.533$; $df = 1$; $p = 0.011$).

Eighty six per cent of the Birchwood respondents were positive about this issue, compared to 73% of the respondents from outside. The reasons for this positive outlook most often cited by the Birchwood respondents were reasons associated with Birchwood's "local green spaces/green setting".

During the interviews the value of Birchwood's green environment for children was explored in more detail. Two main types of benefit were identified: Contact with the natural world; and the possibilities for adventurous play.

The interviews also indicated that Birchwood's green spaces are widely used for family activities and outings.

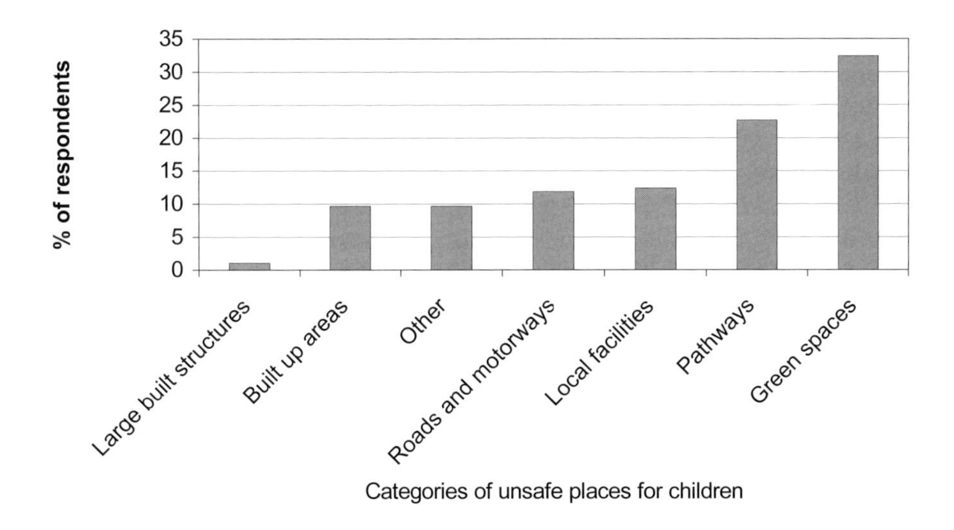

Fig. 8. Respondents' choice of unsafe places for children in the local area

The postal questionnaire also asked the respondents to identify up to three places in the local area where they felt children would be unsafe. The respondents' first-named places were sorted into the same categories used in relation to adults' perception of their own safety, referred to above. "Green spaces" were most commonly identified as unsafe places for children across the whole sample (Fig. 8). It seems that there is a similar contradiction in the way that adults conceptualise their own and their children's' safety in "green spaces". On the one hand many respondents believe these places make Birchwood a good place to bring up children, on the other they are also regarded as unsafe for children to be in.

The interviews suggested that what most adults feared would happen to their children in these places was abduction or assault. In the postal questionnaire the respondents were asked to rank five potential threats in order of the magnitude of the risk they presented to children in the local area.

These threats were: "child abduction/assault", "traffic accident", "bullying", "drugs/alcohol" and "involvement in gangs". Given the finding that "green spaces" were the most commonly identified unsafe places for children and given the interview data suggesting that this was because of the risk of abduction or assault it is surprising that the greatest risk to children in the local area was actually perceived as "traffic accident" (Fig. 9), something that is clearly incompatible with "green spaces". Perhaps this apparent contradiction has to do with the nature of the perceived risk. Although "child abduction/assault" is seen as the smallest risk, it is the most terrifying threat and the fact that "green spaces" are seen as the ideal places for this to happen invests them with a special danger.

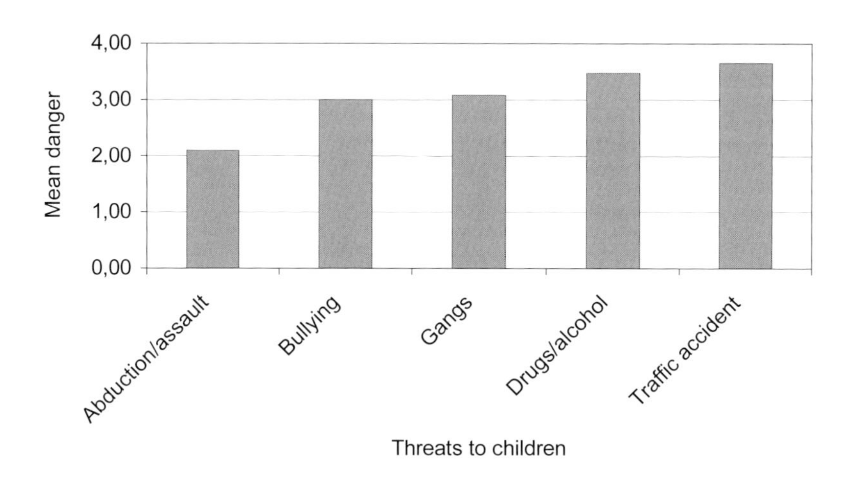

Fig. 9. Respondents' evaluation of dangers to children in the local area

Conclusion

What do these findings tell us about the role of naturalistic woodland in future urban landscapes? The research suggests that such woodland has the potential to be a highly valued part of the urban fabric but that the interface between people's dwellings, the places they frequent daily and the woodland must be carefully handled. This paper suggests a conceptual approach for dealing with this interface. This approach envisages the woodland and the spaces within it as making up three different landscape zones: "the wilderness zone", "the cultivated zone" and "the personalised zone":

- Within "the wilderness zone" users can expect to encounter and interact closely with predominantly nature-like or even wild-looking landscapes, and conserving the integrity of these landscapes will take priority over concessions to user's perception of their personal security.
- In "the cultivated zone" there will be clear signs of human intervention and structure including overtly "maintained" landscapes and formal or ornamental plantings, and the priority will be to maximise user's feelings of personal safety.
- "The personalised zone" will usually consist of residents' own homes and gardens but may also include or overlap with the street, or parts of

it. In "the personalised zone" residents have control over what is planted and or how vegetation is managed.

It is not envisaged that the three zones should be discrete or separate: they can overlap or infiltrate each other. Rather, they are intended as a means of planning and designing with users' needs in mind, and as a means of creating legible landscapes.

It should be emphasized here that this concept is somewhat oversimplified. A full account of the research findings and recommendations are outside the scope of this paper.

What the research clearly showed is that human responses to woodland are extremely diverse. To give but one example of this the research suggested that that many women do not want to enter the woodland in Birchwood on their own confirming previous research in which gender has been found to be very significant in studies of perception of safety in urban landscapes, with women being far more fearful than men (Valentine 1989; Madge 1997; Jorgensen et al. 2002). It is therefore essential for urban landscapes containing naturalistic woodland to incorporate opportunities for individuals to choose whether to interact with the woodland. This has major implications in terms of footpath/transport networks and the strategic location of woodland in relation to places that people need to frequent to perform the activities of daily living (e.g. shops and schools).

Moreover, conventional floral displays are not the only ways of marking gateways or transitions between the so-called "wilderness" and "cultivated zones". There are many ways in which landscape designers can express human care and intervention in the landscape and the challenge is to find innovative ways of evoking familiar responses.

References

Appleton J (1975) The Experience of Landscape. Wiley, London, UK

Bourassa S (1991) The Aesthetics of Landscape. Belhaven Press, London and New York

Burgess J (1995) Growing in confidence – understanding people's perceptions of urban fringe woodlands. Countryside Commission, Cheltenham, UK

Burgess J, Harrison CM, Limb M (1988) People, parks and the urban green: a study of popular meanings and values for open spaces in the city. Urban Studies 25:455–473

Bussey S (1996) Public use, perceptions and preferences for urban woodlands in Redditch. Ph.D. thesis, University of Central England, Birmingham, UK

Department of the Environment, Transport and the Regions (2000) Indices of Deprivation. Department of the Environment, Transport and the Regions, London
Dowse S (1987) Landscape design guidelines for recreational woodlands in the urban fringe. Master's dissertation, University of Manchester, UK
Jorgensen A, Hitchmough J, Calvert T (2002) Woodland spaces and edges: their impact on perception of safety and preference. Landscape and Urban Planning 60(3):135–150
Kaplan R, Kaplan S (1989) The Experience of Nature. A Psychological Perspective. Cambridge University Press, Cambridge, England
Kaplan R, Kaplan S, Ryan RL (1998) With People in Mind. Island Press, Washington DC, USA
Kent M, Coker P (1992) Vegetation description and analysis – a practical approach. Belhaven Press, London, UK
Kowarik I (2005) Wild urban woodlands: Towards a conceptual framework. In: Kowarik I, Körner S (eds) Urban Wild Woodlands. Springer, Berlin Heidelberg, pp 1–32
Madge C (1997) Public parks and the geography of fear. Tijdschrift voor Economische en Sociale Geografie 88(3):237–250
Manning O (1982) New Directions, 3 Designing for man and nature. Landscape Design 140:31–32
Office of the Deputy Prime Minister (2003) Sustainable communities: building for the future. Office of the Deputy Prime Minister, London
Orians GH, Heerwagen JH (1992) Evolved responses to landscapes. In: Barkow JH, Cosmides L, Tooby J (eds) The adapted mind – evolutionary psychology and the generation of culture. Oxford University Press, New York and Oxford
Rohde CLE, Kendle A.D (1994) Human well-being, natural landscapes and wildlife in urban areas. A review. English Nature, Peterborough, England
Tartaglia-Kershaw M (1980) Urban woodlands: their role in daily life. Masters dissertation, Department of Landscape, Sheffield University
Thompson IH (2000) Ecology, community and delight – sources of values in landscape architecture. E and F N Spon, London and New York
Tregay R, Gustavsson R, (1983) Oakwood's new landscape – designing for nature in the residential environment. Sveriges Lantbruksuniversitet and Warrington and Runcorn Development Corporation. Stad och land/Rapport nr 15 Alnarp nr 15.
Valentine G (1989) The geography of women's fear. Area 21(4):385–390
Valentine G (1997) A safe place to grow up? Parenting, perceptions of children's safety and the rural idyll. Journal of Rural Studies 13(2):137–148

Use and Perception of Post-Industrial Urban Landscapes in the Ruhr

Andreas Keil

Institute of Geography and Didactics of Geography, University of Dortmund

Introduction

In post-industrial landscapes, "new wild woodlands" emerge as a result of far-reaching structural changes to regions that were formerly dominated by heavy industry. In the German Ruhr area, this process started earlier and with a greater intensity than in other parts of Germany. Thus, a greater potential for urban-industrial woodlands ("nature of the fourth kind" according to Kowarik 2005) and corresponding activities by local people has arisen in the Ruhr.

To develop innovative strategies for abandoned industrial land was a main goal of the IBA Emscher Park (International Building Exhibition of the Emscher Park). The conceptual approaches were aimed at providing the following functions in the post-industrial urban landscape (see Dettmar and Ganser 1999; Dettmar 2005 on the Project "Industriewald Ruhrgebiet"):

- Compensation function for the balance of nature
- Protection of species and biotopes and creation of a 'preserve' for endangered species
- Political-pedagogical function
- Experience of nature and open space for town dwellers
- Preservation and transformation of industrial-cultural landscapes
- Variety and beauty
- Ecology and aesthetics
- Process observation

The importance of abandoned industrial land as an abiotic and especially as a biotic resource has been demonstrated in several studies (Dettmar 1992; Kowarik 1993; Ganser 1995; Rebele and Dettmar 1996). How-

Kowarik I, Körner S (eds) Wild Urban Woodlands.

ever, less is known about the other functions and their potential for activities of the local population.

Goals and methodological approach

This paper focuses on the question of how local people perceive, adapt and use the new nature that arises on abandoned industrial land, i.e. the relationship between the individual and the place is the subject of the research. This approach is one of qualitative social research; in contrast to a hypothetical-deductive procedure, the research is characterised by processability and reflexivity. Thus the analysis was not aimed at finding the "final truth" or "concrete knowledge", but at gaining knowledge that would provide orientation. Figure 1 shows the different dimensions of the study and illustrates the strong linkage between previous knowledge, empirical procedure and theory formation.

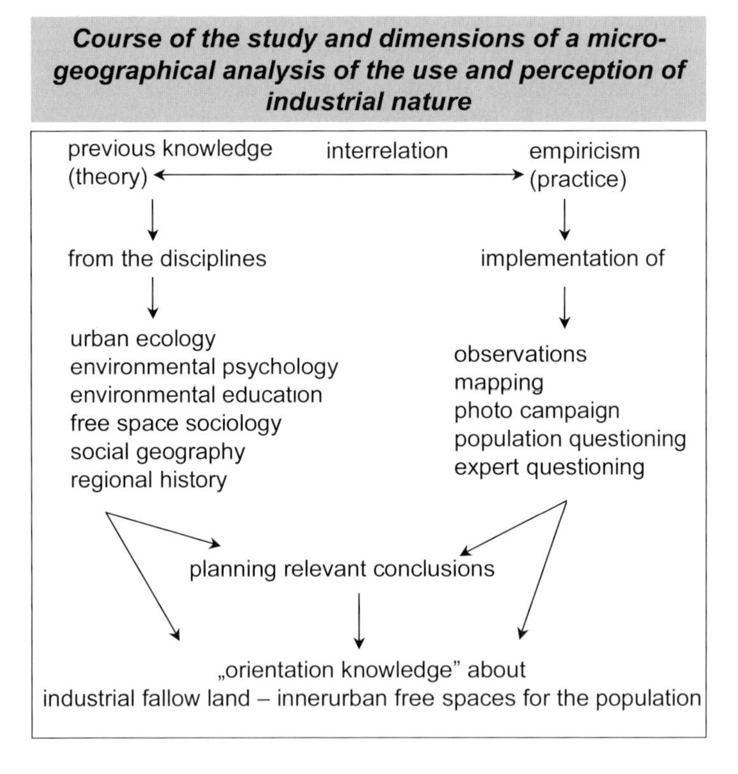

Fig. 1. Methodical concept of the study

The field studies consisted of the following approach (for more details see Keil 1997, 2002; Findel et al. 2003). Observations and mappings were made in the summers of 1997 and 1998. Over 72 days and, in total, 360 hours, the activities of children, adolescents and adults were observed at three post-industrial sites within the Ruhr: "Industriewald Rheinelbe" in Gelsenkirchen, "Landschaftspark Duisburg-Nord" in Duisburg, and "Sinteranlage Ruhrort" in Duisburg. The aim of the study was to collect data on the uses of these post-industrial areas in order to describe and interpret the uses. In the summer of 2003, the survey was repeated by Sibylle Findel at the Rheinelbe area using the same methodological approach (over 24 days, in 120 hours).

To begin to understand people's perceptions of nature, visitors were invited to take photographs of the three areas. Many visitors to the areas participated voluntarily in the photo campaign in order to capture their personal, uninfluenced impressions of the investigated areas. Based on the results of the observations, which are presented in the following sections, and of the photo campaigns, problem-centered interviews were held with 17 users and residents of the analysed areas and with 19 experts.

Adventure sites for children

Pedagogues and developmental psychologists point out that "forbidden spaces" like abandoned industrial land are playing sites preferred by children (e.g. Nolda 1990; Gebhard 1994). They argue that these areas are as fascinating for children today as rubble sites were for post-war children. This "no man's land" represents a "playable environment" (Bochnig and Mayer 1989) and is relatively close to nature.

This study confirms that the numerous structures on industrial fallow land attract many children and young people and stimulate them to different uses (Table 1). Obviously, wild urban woodlands meet a child's need for free, uncontrolled activities and for environments that may be adapted through play.

What do children do in these areas? The normal outside children's games such as ball-catching, hiding, and building tree houses were all observed, but most of the games were specially developed by the children in response to the structure of the area. For example, the games of two ten-year-old boys whom the author met on different sites within the area one afternoon included:

- Climbing the bunker at the east dump

- Repeatedly climbing the east dump and sliding down the slope with cries of joy and pain ("Ouch, seems like Formula 1 here...")
- Catching butterflies and grasshoppers in the meadows
- Racing chariots with old found pushcarts
- Playing with fire on broken-off branches at the east dump
- Exploring the old fire-brigade building

Table 1. Activities in the "Industriewald Rheinelbe" (results of 120 hours of observations and mappings over 24 days carried out in Gelsenkirchen in the summer of 1997)

Activities observed	Number	%
Walking the dog	628	28.1
Taking a walk	467	21.1
Children's play	266	11.9
Cycling	260	11.7
Adolescents' activities	152	6.8
Lingering	100	4.5
Passing through quickly	75	3.4
Mountain biking	74	3.3
Other sport activities	65	3.0
Picking flowers	39	1.7
Collecting berries	30	1.3
Horse-back riding	27	1.2
Jogging	14	0.6
Repairing cars	8	0.4
Making a fire	7	0.3
Catching insects	4	0.2
Gathering mushrooms	3	0.1
Inline-skating	3	0.1
Moto-cross biking	2	0.1
Photography	2	0.1
Shooting games	2	0.1
Hunting	1	0.1
Total	2,229	100

These areas give children a feeling of free movement. They can enjoy their needs without the control and influence of adults. They experience "wildlife" and adventure and feel relatively safe at the same time. Both the freedom of movement and the lack of control are the decisive criteria of these areas for children. Normally they are not aware of the nature that contributes a lot to this wild atmosphere. However, during the games there is unintentional but intense contact and adventure with nature which would not be possible at all, or would be rare at other places in the town. These findings suggest that these areas contribute to the physical and psychological development of children and young people in a positive way.

The use of these areas by children was viewed positively by the adults as well. One of the interviewees, a resident in the vicinity of the Landschaftspark Duisburg-Nord, recalls, misty-eyed, his own childhood:

"They can play and live there in a way which is scarcely possible today. When I think back to my childhood, I remember that there were open spaces close to every residential area where we could romp about, playing all kinds of scouting games like cops and robbers, Indians or the like."

However, in the further course of the interview, the worried father also commented critically on possible dangers.

Previous studies have shown, however, that there are no dangerous risks to children or others from waste on the former industrial areas developed as the IBA Emscher Park. Where necessary, restoration measures were taken. However, dangers potentially arising from the wild and uncontrolled character of the areas should not be removed as this free and rough character allows children to independently adapt and change the areas. Here, urban children can learn to estimate risks and calculate them properly. It should be mentioned that traffic, which is the biggest source of danger for urban children, does not exist in these areas.

In summary, the following thesis for children can be proposed: abandoned industrial land and woodlands in the urban fringe are important inner-urban adventure sites for children. They often have the characteristics and potential of places for experiencing nature (see also Frey 1993; Reidl et al. 2003).

Free spaces for youth

In an urban world widely governed by rules and prohibitions, the wild structures of abandoned industrial land do not only give children the feeling of freedom and adventure. Urban-industrial woodlands like the Rheinelbe area also attract adolescents. The category of "adolescents' activities" in Table 1 summarises activities that teenagers typically pursue in unsupervised spaces. The diverse structures within the woodlands provide places where adolescents can act without being observed by adults. These adolescent activities constituted some of the most frequent activities at the Rheinelbe area during the observation period in 1997 (Table 1). The results obtained from the other analysed areas were similar.

One typical activity of adolescents was the establishment and use of a self-made inline-skating or skateboard run. Located in a clearing surrounded by trees and an earthen bank, it was hardly visible from outside (Fig. 2). This use is not a forbidden one, but at other places the young people would not have been allowed to put up the run without official permission. In addition to the tree houses and huts which appear frequently this is a good example of "self-determined use" or "appropriation activities" as definded by Nohl (1981). The "unfinished" appearance of such an open space stimulates creative activities. One of the remarkable features of the urban woodlands of the Ruhr is that the co-existence of use and nature is allowed.

Fig. 2. Self-built skate run established by adolescents at the Rheinelbe area (summer 1997, photo from photo campaign)

However, a negative aspect must be mentioned. As these activities of young people become excessive, significant vandalism and undesirable "extreme" uses are to be observed as well. In the urban woodlands that belong to the "Industriewald Ruhrgebiet", forestry officers control such activities simply by being present (see Dettmar 2005).

In summary, the following thesis regarding young people can be formed: Industrial fallow land and urban woodlands are important inner-urban rule-free spaces for youth. The nature that has developed on these areas is, however, of little direct relevance to them.

Recreation areas for adults

Even in 1969, social geographers in Germany (Ruppert and Schaffer 1969) believed that easily accessible recreational areas close to a town were more important than the outdoor recreation during a year's holiday (Fliedner 1993). The suitability of open spaces and landscapes for recreation has since been analysed in many studies (e.g. Nohl 1981, 1984). Following this research, open space qualities such as naturalness and diversity, specific characteristics of the site, possibilities for "appropriation activities", accessibility and usability contribute to recreation of adults. The results of this study show that this obviously also holds true for urban-industrial woodlands.

Compared to the survey from 1997 (Table 1), leisure activities of adults increased in 2003 (Table 2) while children's play and adolescents' activities decreased. This shows that the project fields have developed into established and accepted open spaces and have shaped the post-industrial urban landscape of the Ruhr (Fig. 3).

Table 2. Activities in the "Industriewald Rheinelbe" (results of 120 hours of observations and mappings carried out over 24 days in Gelsenkirchen in the summer of 2003)

Activities observed	Number	%
Cycling	766	27.2
Taking a walk	665	23.6
Walking the dog	481	17.1
Lingering	242	8.6
Jogging	229	8.1
Children's play	129	4.6
Mountain biking	118	4.2
Other sport activities	61	2.2
Adolescents' activities	59	2.1
Collecting berries	22	0.8
Passing through quickly	17	0.6
Getting drunk	14	0.5
Picking flowers	10	0.4
Studying industrial history	4	0.1
Photography	2	0.1
Horse-back riding	1	0
Total	2,820	100

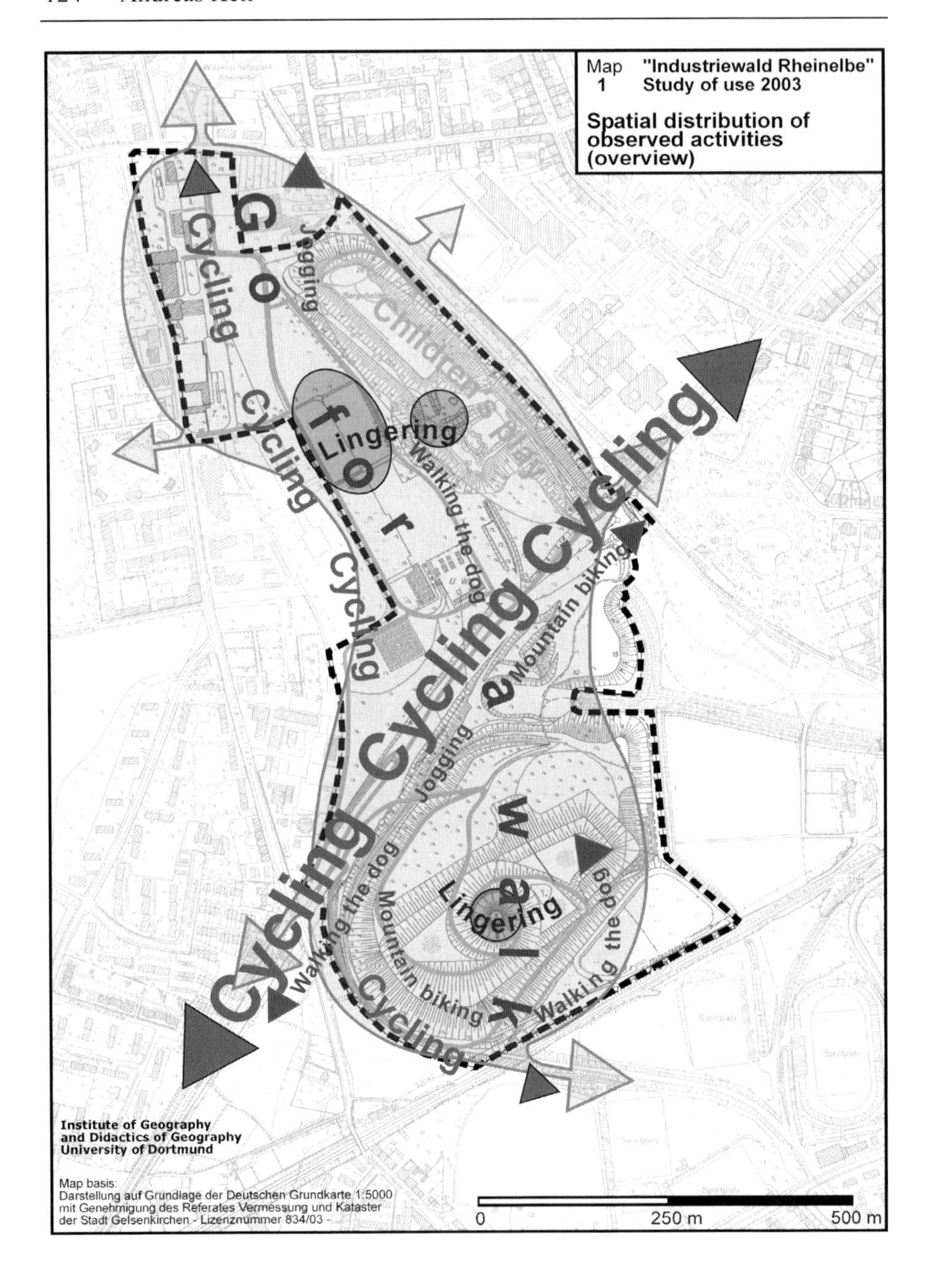

Fig. 3. "Industriewald Rheinelbe" – Spatial distribution of observed activities in 2003

Naturalness and Diversity

The naturalness of abandoned industrial land manifests itself in the spontaneous regeneration and undisturbed development of vegetation over the course of many successive stages up to woodland. The variety of the vegetation of these woodlands is increased by seasonal changes. Moreover the massive, diverse interventions of industry have had a profound effect on the forms of the surfaces, such as excavations, compaction and embankments (dumps, basins), so that the relief of these areas is often clearly structured. This naturalness and diversity in the IBA Emscher Park stands out not only in contrast to urban construction, but especially in contrast to the remaining structures of the former industrial use. Here is a quotation from an interview with a resident of the Landschaftspark Duisburg-Nord:

"How I see the area? Like a sort of paradise garden. Because all the time something new appears, you can observe something different growing at every time, and the colours, it is an enormous gorgeous variety. That is the fascinating thing about it. The transitions of the seasons in spring and the autumn are pure, that is, they are totally extreme and wonderful."

Character

Woodlands on the urban fringe are characterised by the extraordinary and varied juxtaposition of new nature with the remnants of industrial use (buildings, machines, dumps, etc.). Once these elements are understood as aesthetic symbols of the transformation of the area, they catch the attention of the viewer and trigger the demand for more detailed information about the development of the area. Here is another extract from an interview with a resident of the Landschaftspark Duisburg-Nord:

"I think the attraction is that when nature claims something back, it is what we all see and feel, it is not a piece of wood which had always been there before, which we regarded as a damaged and contaminated ground. That somehow counts double for me."

Potential for "appropriation activities"

Old abandoned land which has been fallow for a long time and upon which nature is developing without hindrance can be a place for appropriation activities. These often have an "unfinished" appearance and differ significantly (e.g. in variety, character). The photo shown in Fig. 4 was taken during the photo campaign at the Rheinelbe area. Here, a visitor documents such activities: she lets her dog run free and leaves the path to follow it; she sits down and the meadow thereby becomes "appropriated".

Fig. 4. "Appropriated" meadow on the Rheinelbe area (summer 1997, photo from photo campaign)

Accessibility and usability

Old industrial areas are often in the direct vicinity of residential areas and are connected to them in many ways. However there are still barriers remaining from previous industrial use such as walls and gates, which block access if the owners of the areas do not wish to allow a public use. Intentionally opened areas on abandoned industrial lands such as the IBA projects are freely accessible and equipped with a certain infrastructure for the users. Here too an interview shows how these new inner-urban open spaces are valued:

"The areas are of leisure value, the people get out of their narrow streets, see a bit more of the area and nature" (Interview with a resident of the Landschaftspark Duisburg-Nord).

The aspect "industrial fallow land as leisure space for adults" is to be extented by two dimensions, which result from their past as a place of work: they can show an aesthetic-symbolic importance and influence the regional cultural awareness.

Aesthetic-symbolic importance

The photo campaigns on the abandoned industrial lands within the Ruhr have shown that the aesthetic dimension is perceived as a special characteristic of the area. The photographs taken mirror a variety of "personal views" regarding the industrial fallow land. A frequent theme chosen by the participants was the contrast between "wild" vegetation and remnants of the industrial past (Fig. 5).

Fig. 5. Landschaftspark Duisburg-Nord – "wild" vegetation vs. industrial past (summer 1998, photo from photo campaign)

Landscapes of this kind seem to be especially attractive for the visitors to the area due to the implicitly contained information. In the questioning, it became clear that nobody could resist the aesthetic stimulation of this contrast. When viewing the area, the respondents explained that they were impressed especially by the power of nature and the frailness of human production and that this picture gave them hope for the future.

Regional awareness

Talks with the local population of the IBA project area reveal that at the beginning of the 1990s the industrial remnants were seen as "abandoned polluted areas" and (superfluous) symbols of the economic decline. By the end of the IBA project in 1999, however, this negative image had changed to a positive one. The industrial remnants have increasingly become relics of culture, the fallow lands converted to open spaces shaped by nature. For this change of awareness, a "maturing" time of several years was necessary in order for a certain distance from the previous era to be gained. Only then could the residents of the region open themselves to the new potentials of the area. Quite often they were stimulated by people from outside who had seen the potential before.

Conclusions

This study reveals a generally positive perception of the studied post-industrial areas by different groups of the local population. The results presented have shown that these areas can meet the functions described at the beginning of this text (introduction) for the residents of the region. The post-industrial areas have been used in many ways as additional inner-urban open space that offers the possibility of an active participation with an environment formed by nature in the vicinity of the users' homes. This leads to the following conclusions for further management of post-industrial urban landscapes. Spontaneous development of nature should not be suppressed. Maintenance activities can be reduced to the greatest possible extent. Industrial remnants should be preserved. Their aesthetic-emotional functions are especially high if they can be entered and reused. However, their "conquest" by nature should not be suppressed because this process is especially aesthetically attractive.

Acknowledgements

Thanks are due to the IBA Emscher Park that funded parts of the study in 1997 and to the Ministry of the Environment and Conservation, Agriculture and Consumer Protection of the State of North Rhine-Westphalia that funded the study in 2003. Thanks to Sibylle Findel for carrying out the study in 2003 as well as to Marie-Luise Stein for translating this text into English. I am thankful also to all users and residents of the analysed areas and all experts who took part in our studies and made a major contribution to its successful implementation.

References

Bochnig S, Mayer E (1989) Spielen ohne Spielplatz. Garten + Landschaft 3:25–29

Dettmar J (1992) Industrietypische Flora und Vegetation im Ruhrgebiet (Dissertationes Botanicae 191). J Cramer, Berlin, Stuttgart

Dettmar J (2005) Forests for Shrinking Cities? The Project "Industrial Forests of the Ruhr". In: Kowarik I, Körner S (eds) Urban Wild Woodlands. Springer, Berlin Heidelberg, pp 263–276

Dettmar J, Ganser K (eds) (1999) IndustrieNatur – Ökologie und Gartenkunst im Emscherpark. Ulmer, Stuttgart

Findel S, Keil A, Otto K-H (2003) Industriewald Ruhrgebiet: Nutzung und Wahrnehmung eines neuen Freiraumtyps im Ballungsraum – aktionsräumliche und wahrnehmungsgeographische Untersuchung 2003. Unpublished Research Report, MUNLV NRW, Düsseldorf

Fliedner D (1993) Sozialgeographie (Lehrbuch der allgemeinen Geographie, Bd. 13). De Gruyter, Berlin New York

Frey J (1993) Naturerlebnisräume in der Stadt – Ausgleichsflächen für Menschen und ihre Umwelt. Verhandlungen der Gesellschaft für Ökologie 22:203–209

Ganser K (1995) Landschaftstypen im Emscherraum: Zur Frage ihrer Schutzwürdigkeit. Natur und Landschaft 70 (10):448–453

Gebhard U (1994) Kind und Natur. Die Bedeutung der Natur für die psychische Entwicklung. Westdeutscher Verlag, Opladen

Keil A (1997) Industriebrachen: „Wildnis" im Ruhrgebiet. Empirische Untersuchung zur Ermittlung des Besucherverhaltens, der Besucherhäufigkeit und der Besucherverteilung auf „Restflächen der Industrielandschaft". Unpublished Research Report, IBA Emscher Park, Gelsenkirchen

Keil A (2002) Industriebrachen – Innerstädtische Freiräume für die Bevölkerung. Mikrogeographische Studien zur Ermittlung der Nutzung und Wahrnehmung der neuen Industrienatur in der Emscherregion (Duisburger Geographische Arbeiten, Bd. 24). Dortmunder Vertrieb für Bau- und Planungsliteratur, Dortmund

Kowarik I (1993) Stadtbrachen als Niemandsländer, Naturschutzgebiete oder Gartenkunstwerke der Zukunft? In: Wittig R, Zucchi H (eds) Städtische Brachflächen und ihre Bedeutung aus der Sicht von Ökologie, Umwelterziehung und Planung (Geobot. Kolloq. 9). Natur und Wiss. Hieronimus und Schmidt, Solingen, pp 3–24

Kowarik I (2005) Wild urban woodlands: Towards a conceptual framework. In: Kowarik I, Körner S (eds) Urban Wild Woodlands. Springer, Berlin Heidelberg, pp 1–32

Nohl W (1981) Das Naturschöne im Konzept der städtischen Freiraumplanung. Plädoyer für eine Naturästhetik. Garten + Landschaft 11:885–891

Nohl W (1984) Städtischer Freiraum und Reproduktion der Arbeitskraft. IMU-Institut, München

Nolda U (1990) Stadtbrachen sind Grünflächen. Garten + Landschaft 9:27–32

Reidl K, Schemel H-J, Langer E (2003) Naturerfahrungsräume im städtischen Bereich. Konzeption und erste Ergebnisse eines anwendungsbezogenen Forschungsprojekts. Naturschutz und Landschaftsplanung 35(11):325–332

Rebele F, Dettmar J (1996) Industriebrachen: Ökologie und Management. Ulmer, Stuttgart

Ruppert K, Schaffer F (1969) Zur Konzeption der Sozialgeographie. Geographische Rundschau 21:205–214

People Working for Nature in the Urban Forest

Clive Davies

North East Community Forests

England's community forests

The Community Forest programme in England, announced in 1988, started as an experimental initiative with an ambitious vision for the creation of well-wooded landscapes in and around major urban areas for work, wildlife, recreation and education. Community Forests have evolved into skilled exponents of landscape scale change, developing the contribution of strategic environmental thinking to a wide range of environmental, social and economic agendas.

Community Forests cover large areas around the edges of towns and cities but unlike many forests they are not continuous plantings of trees. Instead, Community Forests comprise a mosaic of wooded landscapes and land uses including farmland, villages, leisure enterprises, nature areas and public open spaces.

They are intended to create areas rich in wildlife, whilst making provision for access, leisure and education - providing attractive areas in which to live, do business and enjoy leisure time.

The Community Forest programme believes that woodland provides a good environment for recreation and can absorb relatively large numbers of visitors without significant loss of visual amenity or damage to nature habitats. Indeed the creation of new Community Forests has led to an increase in the opportunities for creating new wildlife habitats and involving people in this work.

Case study area: the North East of England

The North East of England is the smallest English region, with roughly 4% of the UK's population, land area and economic output. Formerly dominated by energy production, heavy industry and manufacturing, the region

Kowarik I, Körner S (eds) Wild Urban Woodlands.

has seen enormous change over the past 30 years, and is still dealing with the economic, social and structural consequences of its past – including a legacy of derelict and unused urban and peri-urban land.

The North East of England has two designated Community Forest areas, the Tees Forest (set in the valley of the River Tees) and the Great North Forest (covering the lower Tyne and Wear river valleys and north County Durham). Established in 1991 and 1990, respectively, they are highly regarded as successful partnership organisations, involving a total of eleven out of the region's twenty five local government authorities, as well as the national Countryside Agency and Forestry Commission.

The Great North Forest (www.greatnorthforest.org.uk) covers an area of 249 square kilometres, while The Tees Forest (www.teesforest.org.uk) encompasses some 350 square kilometres. These recognised 'brand' names have now been brought together under one heading as North East Community Forests.

Achievements

The two Community Forests in North East England have undertaken a wide range of varied activities in urban and peri-urban areas, helping to create a more attractive and well-wooded environment with accessible and sustainably managed landscapes that enhance the health, well-being and quality of life of local people.

Between 1990 and 2003 the Great North Forest has achieved: more than 800 ha of woodlands planted; over 200 ha of derelict land reclaimed; over 450 hectares of wildlife habitats created or improved.

Between 1991 and 2003 The Tees Forest has achieved: more than 1060 ha of woodlands planted; 342 km of access routes created, upgraded or restored; 290 km of new hedgerows created for biodiversity and landscape improvement.

In addition to physical land-management projects, the two Community Forests have staged a wide range of community and cultural activities, including: art works in the environment; education and out-reach projects aimed at specific communities near to new woodland planting sites; guided walks and health walks; green festivals; events for particular parts of the community (e.g. ornithologists); a series of sporting and active recreation events.

Improvements to access (enhancement of rights of way, transport and information) are also an important part of their work.

Approach to nature projects

The two Community Forests in North East England have undertaken a wide range of nature projects. These are undertaken as an integral part of multi-purpose forest creation or forest management. This implies that nature activities are undertaken within a wider range of crosscutting themes, which includes leisure, recreation, sport and landscape improvements. The balance of these crosscutting themes is established at the design stage of a project and the scheme is then designed around them. For example in some cases a high priority is given to nature values and this may put tight limits on recreation opportunities and visa versa.

In most cases new projects have a strong measure of public involvement, especially on sites that are close to residential areas. Nature activities have proven popular with local people and are encouraged by Community Forest staff as they are of a scale where public involvement can yield quick results.

The Community Forests employ staff who are trained as Community Liaison Officers. These staff have expertise in nature management as well as in community development. In the North East region this is further strengthened by the employment of a biodiversity specialist.

Case studies from North East England

Conservation of the Black Poplar tree, *Populus nigra*

Populus nigra was once a common tree of the English countryside, but is now rare. It is recognised as an important conservation species in national and regional biodiversity action plans. A regionally significant population of *Populus nigra* is located within the municipality of Darlington and within the boundaries of The Tees (Community) Forest.

Objectives. Raise local public awareness of the tree and its conservation needs, establish community run tree nurseries to propagate new trees from local sources and plant back into urban and peri-urban locations using intermediate labour market employees (training for employment and volunteers) with the agreement of local land managers and farmers.

Key stakeholders. Darlington Borough Council, The Tees Forest, Biodiversity Action Plan, Durham Wildlife Trust, Local Heritage Initiative.

Resources. Financial support from Local Heritage Initiative (national lottery funding); staff contributions from Darlington Borough Council, NECF

Land Services and The Tees Forest, participation by stake-holders, local volunteers and school children.

Actions undertaken. Public awareness events and festivals, work in schools by community liaison officers, national and local media coverage including the BBC, music CD produced for promotional purposes.

Black poplar tree nurseries established in local schools, planting of trees on sites to increase population and distribution, appointment of a 'Black Poplar Project Officer', development of a Black Poplar tree nursery.

Number of people involved. 2,000.

Fig. 1. Sculptured seat provides a viewing point for a *Populus nigra* tree prominent in the local landscape

Outcomes. Increased public awareness; involvement in a conservation issue of local significance; increased public understanding of the Black poplar and its historical significance, increased public awareness of the stakeholder organisations, increased local biodiversity. The value of the carbon sequestered by this project and importance of the biodiversity to the local population has been calculated at €30,000 per annum.

Evaluation. A successful project, which has received good media coverage, support from local decision makers and involved a significant number of local people.

Stainton Quarry local nature reserve

Stainton Quarry is located within an attractive and prosperous village environment on the southern outskirts of the town of Middlesbrough. The Quarry has naturally regenerated to mature woodland but includes other habitats including species rich grassland. The area is used extensively by local people for recreation, of which exercising dogs is the most popular. At times the site has suffered from vandalism and this has raised concerns about the use of the site from local residents. Local residents help manage the site.

Fig. 2. Panoramic view of Stainton Quarry local nature reserve, showing that local housing overlooks this green area

Objectives. To secure the biodiversity content of the site and to raise awareness of the wildlife the site contains amongst local residents. Achieve local nature reserve status for Stainton Quarry. Improvements to be made to recreational infrastructure following consultation with residents, such as paths, which will provide for easy access for all ages and mobility. Overcome vandalism problems.
Key stakeholders. Middlesbrough Council, The Tees Forest, Friends of Stainton Quarry, English Nature.
Resources. Staff input from The Tees Forest and Middlesbrough Council. Funding from these organisations and English Nature. Volunteer input from the friends of Stainton Quarry.
Actions undertaken. Management Plan produced outlining how the site will be managed with local residents input; community involvement activities in woodland management; daily community wardening with aim to reduce vandalism; encouragement of wildlife through for example the siting of nest and roosting boxes.
Number of people involved. 50.

Outcomes. Good community participation but local issues over the amount of access to be allowed; vandalism has been reduced; stronger level of 'ownership' by local community than hitherto; involvement of local municipality; Local nature reserve designation achieved and funding attracted from English Nature.
Evaluation. A successful project which has built a strong commitment from local residents to manage wildlife within their near environment, support from local decision makers has been achieved and significant number of local people have been involved either through consultation or by direct volunteering activities.

Stillington Forest Park

The transformation of a former industrial slagheap and tip into a 7 hectare Forest Park in 1994 is the result of the work of a regeneration partnership involving the local municipality, governmental agencies and non-governmental bodies including the Tees Valley Wildlife Trust and The Tees Forest partnership. Extensive use of sub soils in the regeneration scheme has resulted in a highly species rich flora and invertebrate population in addition to new woodland cover.
Objectives. Manage the biodiversity interest of the Forest Park and retain its biodiversity. Build links with the local community and utilise the site for public events. Secure local nature reserve status and funding for the long-term management of the site. Reduce litter problems associated with fishing the lake.
Key stakeholders. Stockton-on-Tees Borough Council, The Tees Forest, Tees Valley Wildlife Trust, BTCV, Local primary school, English Partnerships, Forestry Commission, Countryside Agency.
Resources. Staff input (one and a half days a week) from Stockton-on-Tees Borough Council Countryside Warden services (including volunteers). Financial support from the local municipality and English Partnerships. Input from a Prince's Trust volunteer towards site maintenance.
Actions undertaken. Management Plan produced, appointment of a part-time countryside warden, regular meeting with the local Parish Council, involvement of local school children and development of an outdoor study area, sympathetic management of the local flora, fauna and invertebrates, holiday activities for local children, establishing a fishing club and volunteers group, construction of a permanent orienteering course, venue for an annual trail race, survey of local residents to establish their views of the site.

Fig. 3. Reclaimed 'slagheap' at Stillington has been converted into a developing Forest Park. The site is now famous for its floral diversity.

Number of people involved. 200.
Outcomes. Use of the site by local residents is increasing, biodiversity has been secured and long term education benefits achieved through liaison with the adjacent primary school.
Evaluation. The inter-relationship with the local community has taken longer to establish than on other projects in the local area, the appointment of a part-time countryside warden is seen as a turning point in achieving this. Increasing use of the Forest Park and setting up a fishing club appears to be helping reduce a litter problem.

Woodland wildflower project

Woodland wildflowers have declined in the last 100 years as a consequence of poor woodland management and habitat loss through built development. The creation of new 'community forests' enables initiatives to be run that will restore woodland wildflowers to existing woodland and new woodland sites.

Most of the new planting sites in Community Forests are on existing arable agricultural land, improved grasslands or on reclaimed former industrial sites. Frequently these lack a local seed source resulting in the need for intervention in the form of seed collection, germination, growing on of young plants and subsequent planting.

The Woodland Wildflower Project is a joint initiative of 'LandLife' and England's Community Forests with the participation of the Local Heritage Initiative and charitable trusts.

Fig. 4. School children collecting wildflower seeds

Objectives. Collection of viable wildflower seed from old woodland sites. Involvement of local community groups and school children in the collection of the seeds. Sending viable wildflower seed to the 'LandLife' nursery in North West England for germination and growing on. Planting of young wildflower plants on new woodland sites. Local publicity for the project and increased awareness of woodland wildflowers.
Key Stakeholders. Local authorities, The Tees Forest, Great North Forest, Landlife (Conservation charity), Local Heritage Initiative, National Community Forest partnership, Countryside Agency.
Resources. Finance from the Great North Forest and The Tees Forest, Local Heritage Initiative, Charitable Trusts. Staff resources from The Tees Forest and the Great North Forest. School children and community groups collecting seeds.
Actions undertaken. Seed collection events, growing on of seeds into young plants for transplanting, seed sales, work in schools by community liaison officers, local media coverage, 'totem' species used for marketing e.g. *Silene dioca* in The Tees Forest.
Number of people involved. 2,000.
Outcomes. Tree focused organisations focus on a broader conservation issue, increased public understanding of woodland wildflowers and the reasons behind their decline, relatively poor media response, biodiversity gains localised.
Evaluation. The project was complicated to run suggesting that a simpler local scheme would have achieved greater outcomes and lower cost. North East Community Forests already run a tree nursery in the region and the potential of this being diversified to include wild flower propagation is underway. The number of plants placed back into new woodlands is small and will have little impact in terms of increasing biodiversity.

Coatham Wood Community Forum

Coatham Wood is a new peri-urban woodland established on former arable land (200 hectares) adjacent to an established village community. The stake-holders were concerned about local attitudes to such a major change to the landscape and keen to involve local people in terms of consultation on woodland design and to encourage use of the site and conservation activities.

Fig. 5. Local residents are already making use of this major new woodland

Objectives. Establish a community forum and secure broad support for the proposal to create a new community woodland. Design the new woodland to meet local community aspirations and address concerns at the outset of the project. Maximise habitat creation potential of the site whilst meeting other local demands.
Key stakeholders. Forest Enterprise, The Tees Forest, Stockton-on-Tees Borough Council, Long Newton Parish Council, Primary School, Butterwick Hospice, Admiralty Site Ecology Group.
Resources. Staff and financial resources from The Tees Forest and Forest Enterprise.
Actions undertaken. Consultation meetings, production of community newsletters, tree planting events, nature walks guider by a Forest Ranger, wildflower planting, community orchard proposed.
Number of people involved. 500.

Outcomes. Community Forum successfully established and no significant opposition for the proposals, trees and woodland wildflowers planted by local residents, involved local school children, engaged state forest service in creating and managing the site.
Evaluation. A successful approach endorses the belief by stakeholders that involvement of local residents is essential before site works commence. Fringe benefits include 'active citizens' supporting the new woodland within the local community.

Tool kit for involving people in nature projects

A tool kit on how to involve local people based on the Community Forest experience is set out below. This is based on the situation in England and changes may be needed to adapt it to different cultural situations. The tool kit is set out as a fictional illustration: a proposed species conservation project.

Vision

The organisation seeking to engage people in nature projects should have a clear vision of the project and how people can participate.
Example. Our vision is to conserve the (name of species) in the local area and provide new habitats where the species can flourish, we will do this by collecting seeds, propagating them in local community based nurseries and planting young plants into sites in the local area. People will be involved in collecting seeds, planting and tendering them in the nursery and planting them on sites. People will be able to help the project indirectly by donating money and helping to publicise it.

Publicity

Publicity can raise awareness in the local community as well as educate individuals and secure the support of politicians and decision makers. Good publicity also provides a 'feel good' factor amongst stakeholders.
Example. We will publicise the project by issuing press releases and making direct contact with journalists. We will invite local people to share news of the project with their friends and acquaintances. The project will be recorded photographically and these placed on a website. Events will be organised at convenient times (for instance after working hours) and these

will be targeted at sections of the community. The services of a local celebratory could be sought.

Staff

Staff input (or that of experienced volunteers) is frequently necessary to start a project. Staff time is expensive and limited hence a major objective is to make the project as self-sustaining as possible and withdraw staff time as the confidence of local volunteers increase.
Example. We will dedicate (amount of staff time) at the outset of the project and aim to reduce this by (X)% after (number) months. We will provide free training to local volunteers on how to sustain their activities as staff time decreases and provide free 'help-line' telephone numbers if they encounter unexpected difficulties.

Finance

All projects have a cost. The involvement of volunteers can significantly reduce cost as their time is given for free. A budget is needed to pay for local advertising and costs associated with printing promotional and educational materials. Where a local community group is involved a grant can empower the group to buy in services and reduce the administrative demands on the sponsoring organisation. Community groups are expert in securing help-in-kind (such as low cost transport and machinery).
Example. A budget of (amount of money) will be available to support this project and a proportion of this will be paid to the group to enable them to promote the project. They will provide an annual report on how the money has been spent.

Partnership

Local people may already be involved with other local groups and initiatives. Building links with these groups can increase the opportunities for participation and result in more resources.
Example. We will identify all the local groups and organisations likely to be interested in this area of work (i.e. market research) before launching the project. We wish to involve as many of these bodies as stakeholders and are prepared to share the publicity opportunities with them.

Evaluation

Project evaluation ensures that future projects learn from previous experience. Evaluation gives confidence to stakeholders and investors that the organisation has good project management. Local people value the results of the evaluation and their role in the success of the venture. Evaluation provides a reason to 'thank' people for their involvement.
Example. We will evaluate the project by seeking feedback from all participants throughout the life of the project. A final report will be produced which includes the results of the evaluation. This will report what has been achieved (including photographs), how resources were spent, copies of press articles, comments from key volunteers and members of the public.

Acknowledgements

To the following for their involvement in the Case Studies; Simon Blenkinsop and Sarah Ransome of The Tees Forest, Stockton-on-Tees Borough Council Countryside Wardens, Rob George of Darlington Borough Council.

Nature Returns to Abandoned Industrial Land: Monitoring Succession in Urban-Industrial Woodlands in the German Ruhr

Joachim Weiss, Wolfgang Burghardt, Peter Gausmann, Rita Haag, Henning Haeupler, Michael Hamann, Bertram Leder, Annette Schulte, Ingrid Stempelmann

Introduction

The considerable decline of heavy industry (coal, steel) has led to extensive structural changes in the German Ruhr, leaving behind areas which have been shaped by the profound impacts of industrial use. As long as these industrial abandoned lands are not subjected to use, they experience ecological succession. In 1992, more than 8,000 ha of such land were known in the Ruhr (Tara and Zimmermann 1997). Today the area is estimated to total approximately 10,000 ha.

Depending on the local economic situation, some of these industrial abandoned areas will be used again, for example, as commercial districts, office buildings, housing estates, or as traditional parks. On other sites, however, undirected natural succession continues and produces a mixture of different successional stages that are dominated by short-lived pioneer species, tall herbs, shrubs and trees. On many areas, succession has already led to urban-industrial woodlands which differ in many ways from other types of urban forests (Kowarik 2005). These areas provide considerable potential for developing new green spaces with significant social and ecological functions.

Since 1996, 12 urban-industrial woodlands with a total of 244 ha have been included in the *Projekt Industriewald Ruhrgebiet* (Industrial Forest Project of the Ruhr). This project aims to develop these areas by integrating social and ecological goals (Dettmar 2005). Both cultural remnants and natural processes are found to be highly attractive to local residents (Keil 2005), and the urban-industrial habitats clearly harbour a high number of animal and plant species, including rare species (e.g. Rebele and Dettmar 1996; Gausmann et al. 2004).

Due to the newness of these anthropogenic sites, the way in which ecosystem development will proceed, however, remains an open question.

Kowarik I, Körner S (eds) Wild Urban Woodlands.

This paper reports on a monitoring approach established in 1999 and presents results of the first analysis.

The monitoring approach

The monitoring approach integrates the studies of soil, vegetation and selected animal groups. In total, six permanent plots representing different stages of succession were established. Two plots each were set up at the pioneer, the shrub and the woodland stages.

The heterogeneity within the plots does not allow generalisations regarding succession on industrial abandoned land. Instead, the monitoring aims to document, through examples, the development of and mutual interactions between soil, vegetation and fauna during succession from pioneer to late-successional stages of industrial woodlands. According to the “space-for-time" method, a comparison of plots at different stages of succession will allow tendencies in successional trends to be described (Pickett 1989, Nobis 1998, Purtauf et al. 2004).

The monitoring program is funded by the Ökologie Programm Emscher Lippe (ÖPEL; Ecology Program of Emscher Lippe) and the State Agency for Ecology (LÖBF). The LÖBF co-ordinates the work which is carried out by the University of Essen, the State Agency for Environment, LUA (soil analysis), experts of the LÖBF, the University of Bochum, and independent researchers (on vegetation dynamics, selected animal groups).

Study sites

Three urban-industrial sites were chosen for monitoring the development of soils and the dynamics of plant and animal populations. All study sites are located within the core of the urban-industrial agglomeration of the Ruhr. All were used formerly as coal mines and are now integrated into the Projekt Industriewald Ruhrgebiet, which ensures the future existence of the sites. The first is Zollverein coal mine (20 ha), located within the city of Essen. The other sites are within the city of Gelsenkirchen: the Rheinelbe coal mine (42 ha) and the Alma coal mine (26 ha). The area of each plot is 0.1 ha, including a subplot for vegetation relevées (100 m^2). The other analyses of vegetation structure, soil profiles and faunistic sampling were done outside of this subplot. Table 1 gives further information on the studied plots. For five plots, the parent material for soil genesis was hard coal-mining spoil from a depth of more than 1,000 m; the ecosystem dynamics on the remaining plot started with building rubble.

Table 1. Location of the permanent plots (0.1 ha) on three sites (former Alma, Rheinelbe and Zollverein coal mines). The ecological features describe site and vegetation characteristics at the beginning of the investigation in 1999

Successional stage	Alma (A)	Rheinelbe (R)	Zollverein (Z)
Pioneer stage (P)	Plot PA: recent hard coal-mining spoil with bare ground		Plot PZ: hard coal-mining spoil with sparse vegetation
Shrub stage (S)	Plot SA: rubble with tall herbs and shrubs	Plot SR: hard coal-mining spoil with 5- to 10-year-old birch stand	
Woodland stage (W)		Plot WR: mining spoil dominated by 40- to 50-year-old birch	Plot WZ: mining spoil dominated by 80- to 90-year-old planted black locust

Table 2. Methodological approaches for analysing chemical, physical, microbiological and zoological soil properties in the soil profiles with information on the planned repetitions (repetition intervals in brackets)

Parameters	Method
Chemical (10 years)	
pH value	In 0.01 M $CaCl_2$
Electric conductivity	Electrode in 1:5 soil-to-water suspension
Carbonate content	Volumetric CO_2 measurement with Scheibler-Finkener alkalimeter
Cation exchange capacity (CEC)	Method of Mehlich (Schlichting et al. 1995)
Plant-available phosphate and potassium	By VDLUFA method (Hofmann 1991)
Plant-available magnesium	By calcium-chloride extraction
Content of heavy metals	Aqua regia-extractable contents of Cd, Zn, Pb, Cu, Ni
Pedogenous iron and manganese oxides	Extraction with dithionit (Mehra and Jackson 1960); extraction with NH_4 oxalate according to Tamm/Schwertmann (Schlichting et al. 1995)
Physical (10 years)	
Soil-water content	Gravimetric

Table 2. (cont.)

Bulk density	Gravimetric measurement of 500-cm^3 soil samples
Humus content	Combustion residue loss (gravimetric)
Microbiological (3 years)	
Soil respiration	Oxygen uptake, a Sapromat (Schinner et al. 1993)
Substrate-induced respiration	Oxygen uptake after addition of glucose, Sapromat (Schinner et al. 1993)
Microbial biomass	Indirect estimation by conversion factor from substrate-induced respiration at 22°C (Anderson and Domsch 1978; Alef 1991)
Dehydrogenase activity	With the substrate TTC incubated for 24 h; modified from Thalmann (Schinner et al. 1993)
Cellulose degradation	Litterbag method (Bocock and Gilbert 1957; Dunger and Fiedler 1989; Alef 1991)
Zoological (5 years)	
Lumbricidae: species richness and abundance	formalin expulsion/hand picking
Enchytraeidae: species richness and abundance	Wet extraction according (Dunger and Fiedler 1989)
Springtails (Collembola): determination to genus level and abundance	Dry extraction by Berlese-Tullgren method

Methods

The methods for analysing chemical, physical, microbiological and zoological soil properties are summarised in Table 2. At each permanent plot, a soil profile is dug and analysed according to Hoffmann (1991). Soil types are characterised following the standardised methods (AG Boden 1994, AG Stadtböden 1997). Microbiological parameters are analysed in mixed soil samples.

Vegetation is sampled in 10x10 m^2 plots. Each has four subplots of 1x1 m^2. Cover of plant species is estimated in absolute percentages. In the subplots, individuals of all plant species are counted. Special funnel traps were designed for capturing the diaspore rain (Fig. 1). They are emptied every two weeks. The diaspore bank of the top soil layer is sampled at depths of 0–2, 2–5 and 5–10 cm using a metal cylinder (500 cm^3). Half of the material is studied using the “seed washing by sieve” method, the other half using the “seed emergence” method according to Fischer (1987). For the latter, diaspores are sown on a sterilised soil matrix and observed for a four-week period.

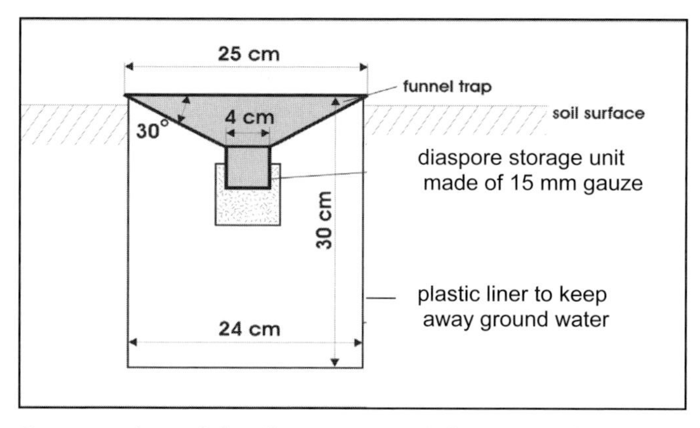

Fig. 1. Construction of the diaspore trap (after Haeupler et al. 2003)

On each permanent plot, the locations of all trees are established to create structure maps of the stand (Leder and Leonhardt 2003). Additionally, the height of the trees, their diameter at breast height (dbh) and their social positions are recorded, using the classification method of Kraft (1884). This classification describes the rank of a tree in comparison to its neighbours by recording the form of its crown. The classes are: (1) superior, (2) dominant, (3) co-dominant, (4) intermediate, and (5) overtopped. For the demographic analysis, saplings with a height of more than 1 m are considered.

By digging up the roots, information about mycorrhiza and root growth is obtained. The analysis will be completed by documenting historical development, and by analysing annual rings and aerial photographs.

The permanent plots are sampled for wild bees, sphecids, hoverflies (syrphids), ants and ground beetles. These groups are chosen because of their suitability as bio-indicators. The ecological requirements of the groups are well known, their lifestyles are adapted to all studied successional stages, with differing habitat requirements within each group. They can be assigned to different trophic levels and use more than one habitat type or habitat structure during their life cycle.

In this paper, only ground beetles and ants are considered. For sampling ground beetles, pitfall traps are used (6 traps per plot; standing time 3x3 weeks per year). In addition, some beetles are caught manually at random. Ant nests are mapped in two 50 m^2 partial areas per sample plot. The ant studies are completed by manual catches within the total sample plot, in pitfall traps used for ground-beetle mapping, using litter-layer sieving in woodland plots, net catches within higher vegetation and manual catches at random. Due to annual fluctuations in populations, samplings of two consecutive years are evaluated as one investigation unit with a regular repetition every five years. The initial recording was carried out in 2000/2001. An additional investigation was carried out in 2003.

Results

Soil genesis and chemical characteristics

Hard coal-mining spoil forms the parent material for soil genesis on five permanent plots. Differences among these sites are mainly due to varying gravel and stone (skeleton) contents. With its strong compacted layers of brick debris and a mixture of building rubble, the plot SA differs greatly from all others.

The soil classified as "skeleton-humus soil" on the pioneer sites PA and PZ consists of coarse mining spoil over loess with initial humus accumulation in cavities. A humus topsoil has not developed yet. Mining spoil of younger fills is often compacted. Shallow stagnic soils develop which have temporary wet topsoils. This is a characteristic feature of site PA.

On the shrub sites, a Syrosem-Pararendzina from building rubble over loess was found at site SA and a Syrosem-Regosol from hard coal-mining spoil over loess was found at site SR. "Syrosem" indicates an incipient soil development of raw soils. Due to the rapid soil development by humus accumulation in the humid climate, the soils are already in transition stages to Pararendzina and Regosol from material containing or lacking carbonates respectively. Different thicknesses of the humus horizons (Ah) have been recorded for the two areas.

The two woodland locations show Syrosem and Regosol of hard coal-mining spoil over loess (WR) or a Regosol of hard coal-mining spoil over loess (WZ). The two sites differ mainly in their stage of soil development. The soil development has not yet proceeded on WR due to erosion on its steep slopes. The development of a humus topsoil has been greatly hindered by this. Only small areas of vegetation provide for local soil accumulation and therefore conditions for humus formation. The site WZ, in contrast, is located in a flat area and 20-cm-thick organic layers (L, Of, Oh) and topsoil horizons (Ah) have developed in places. Thick, raw humus layers indicate an acid soil. Old fills of hard coal-mining spoil, in contrast to the younger ones, are usually loosely packed. Site WZ provides a good example of uncompacted soils developing from hard coal-mining spoil.

The investigated soils are at the beginning of their development. They are very gravelly and stony (skeletic). Down to a depth of more than 1 m, the parent material frequently consists of more than 80% skeleton. Therefore the initial stages of soil development on hard coal-mining spoil strongly restrict root growth. Weathering of skeleton to fine-earth particles is in an initial stage. Therefore the fine-earth content is low. Fine earth supplies plants with nutrients and water. Due to the limited fine earth content only small amounts of nutrients and water can be stored and are, therefore, less available for plants. The nutrient status must be classified as poor with regard to magnesium, phosphate and potassium, except for site WZ. The low fine-earth content also reduces the cation exchange capacity, which is usually very low. The pH values (Fig. 2) on old hard coal-mining spoil are mostly very low (less than pH 4.0; Burghardt 1989). Hard coal-mining spoil that is only a few years old and building rubble material, however, have a moderate alkaline pH value of 7.1–9.0.

Salt washouts can be traced in the decline of electrical conductivity of the soil-water suspension after 1 year at site PA. In 1999, we measured approximately 290–330 µS/cm in the upper 10 cm layer; 1 year later the electrical conductivity dropped to approximately 250–260 µS/cm. This is the "advance notice" of acidification, which usually starts immediately after the washout of the salts.

The heavy-metal concentrations of the fine-earth fraction are mostly slightly elevated. In the raw humus layer at site WZ, 116 mg of copper per kg were measured. Levels of zinc above 150 mg/kg can often be measured in the topsoil. Lead content frequently lies between 76 and 552 mg/kg in the fine-earth material. Nickel and cadmium, however, do not show elevated results. It is striking that plant-available and water-soluble (ammonium-nitrate extract) lead occurs at slightly elevated levels of more than 0.1 mg/kg.

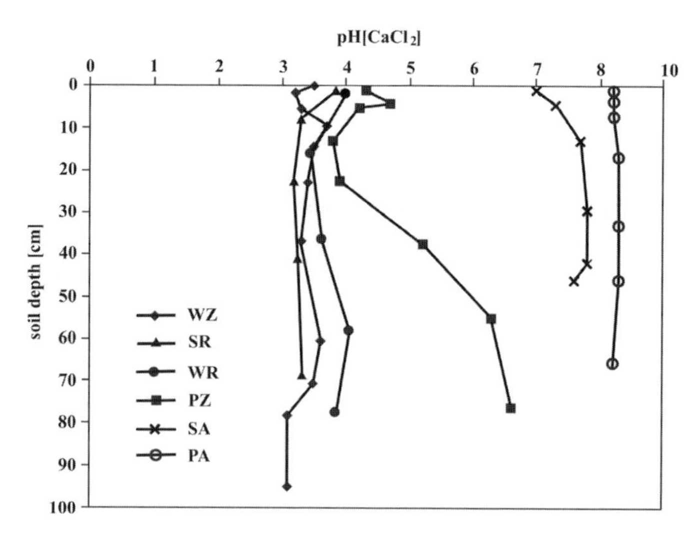

Fig. 2. Depth profiles of the pH values at the sampled plots (for abbreviations see Table 1; after Burghardt et al. 2003)

Soil fauna and microbiological activity

Few organisms of the macro- and mesofauna live in the raw soils. The pioneer sites are still free of earthworms. Generally, no microbiological activity was found on these sites (average soil respiration: 0.04 mg O_2/100 g dry matter/h). In the shrub and woodland sites, the level of microbiological activity ranged between mean moderate and high (average soil respiration from 0.1–1.2 mg O_2/100 g TS/h). Due to ongoing acidification, a reduction in microbiotic activity is to be expected. This would enhance fungi instead of bacteria and would subsequently decrease the decomposition rate of litter.

In the shrub site SR and woodland site WZ, the population density of lumbricids is still low with 6–12 individuals per square meter (Fig. 3). The lack of litter consumers, particularly big, deep-digging earthworms, leads to a "decomposition jam" and an accumulation of litter and thick raw humus forms. Such raw humus forms are typical of the acidic, low-nutrient soils that are found at the site WZ. The lumbricid presence at WR is completely different. Here, big, deep-digging earthworms occur abundantly throughout the mineral soil although they live under the same acidic conditions as the previous cases ($pH_{CaCl2}<4$). In contrast to the lumbricids, the enchytraeids indicate a future change in the composition of the soil macrofauna in response to acidic soil conditions. The dominant species *Cognettia sphagnetorum* and *Marionina clavata* are typical micro-annelids of acidic top layers.

At the rubble site SA, the number of lumbricids is typical of moderately alkaline pH values. Surprisingly, there is a clear dominance of animals that usually live in humus layers over those living in mineral-soil layers, although we recorded no significant humus layer at this site.

Dynamics of diaspore banks

The diaspore bank has been analysed at both shrub sites. In three years of investigation, 58 plant taxa were found in the diaspore bank at the rubble site SA (see Appendix, Table A1). The above-ground vegetation of this site is dominated by tall herbaceous perennial plants such as *Solidago gigantea* and some woody species such as *Salix alba* and *Populus alba* with a total of 48 species recorded in the 10x10 m^2 plot. The results illustrate the changes in the diaspore banks. However, fluctuations in the occurring taxa also depend partly on the chosen study method. Not all taxa found during seed washing left viable seeds (e.g. *Centaurium pulchellum,*

Chenopodium rubrum), and applying the seed-emergence method revealed a group of additional taxa (e.g. *Echium vulgare, Festuca rubra* agg.).

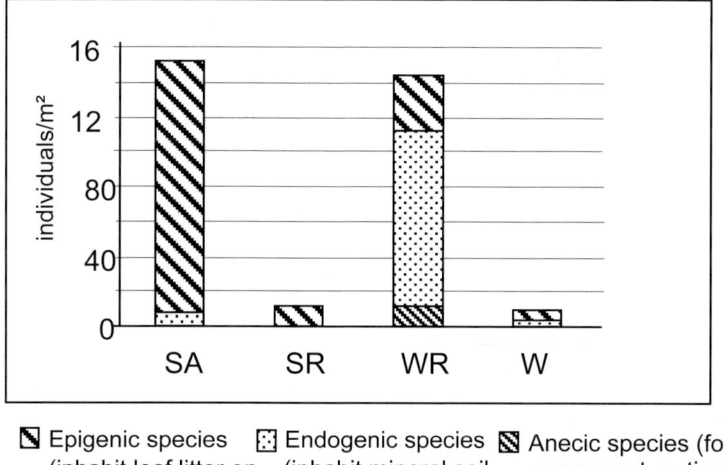

Fig. 3. Abundance and relative proportion of life forms of the lumbricids at the sampled plots. Only shrub and woodland sites are shown, because the pioneer sites were free of lumbricids (for abbreviations see Table 1)

The changes in the diaspore bank diminish further with the progression of ecological succession. Therefore, not only is the number of taxa in the woodland plots low, but the variation in the woodland taxa over the years is extremely low as well.

The other shrub site, SR, is already dominated by pioneer trees (*Salix caprea, Betula pendula*). Similar to the previous site, birch is the most common species in the diaspore bank, due to a high propagule pressure at both sites. However, at SR, only 25 taxa were found in the diaspore bank (Appendix, Table A2).

Demographic structure of woodland stages

The demographic structure of the populations of woody plants was analysed at both woodland sites (Table 3). The site WR is an approximately 40-year-old stand that is still dominated by European birch (*Betula pendula*) as a pioneer species. Some late-successional species indicate the future direction of succession. Most common is the shade-tolerant sycamore maple (*Acer pseudoplatanus*) which is already established under the canopy of birch (Fig. 4). Other tree species occur in the herb layer (*Acer campestre, A. platanoides*, *Carpinus betulus*, *Crataegus* x *macrocarpa*, *Prunus serotina*, *Quercus robur*). In the herb layer, typical early-successional species are still present (*Agrostis stolonifera*, *Cerastium holosteoides*, *Epilobium ciliatum*), but woodland species indicate the further development of the stand (*Dryopteris filix-mas*, *D. carthusiana*, *Deschampsia flexuosa*). The moss *Mnium hornum* is indicative of the strongly acidic soil conditions.

At the second woodland plot (WZ), the North American black locust (*Robinia pseudoacacia*) had been used for recultivation purposes. Within the 80- to 90-year-old stand, a strong natural dynamic is apparent. The black locust is still dominant, but no increase in height occurs in diameter classes above 30 cm (Fig. 5). Instead, some trees are already beginning to die. Gap formation enhances the regeneration of both black locust and other trees. Black locust still predominates in the shrub layer. Due to the poor decomposition of its litter, a thick raw humus layer has accumulated. The herb layer mainly consists of *Rubus* spp. (*R. elegantispinosus, R. nemorosoides*) and fern species (*Dryopteris dilatata, Athyrium filix-femina*).

Table 3. Demographic analysis of urban-industrial woodlands. *Site WR* 40-year-old stand dominated by birch, *site WZ* 80-year-old stand dominated by *Robinia*. Only saplings with a height of more than 1 m are considered. *K 1* refers to the Kraft classification "superior" trees, *K > 4* to "overtopped" trees, *asterisks* indicate data that were not analysed

	Saplings / trees per hectare		Diameter at breast height (cm)			Height (m)			K 1 (%)	K<4 (%)
	n	%	mean	max	min	mean	max	min		
Site WR										
Betula pendula	2610	60	9.6	21.8	0.9	10.1	17.5	1.5	28	14
Sambucus nigra	960	22	3.7	8.4	2.0	3.6	3.7	3.6	*	49
Acer pseudo-platanus	620	14	3.9	16.7	0.4	3.6	9.3	1.4	2	16
Prunus padus	30	1	3.8	4.9	3.1	*	*	*	*	*
Fraxinus excelsior	30	1	2.8	3.3	2.3	*	*	*	*	*
Salix caprea	30	1	19.1	27.0	6.3	*	*	*	67	33
Salix cinerea	30	1	10.1	12.7	7.4	8.2	11.5	6.1	*	33

Table 3. (cont.)

Site WZ										
Robinia pseudoacacia	840	77	21.9	52.5	2.2	7.9	21.5	1.3	6	64
Sambucus nigra	220	20	4.2	7.8	2.1	3.2	4.5	1.0	0	100
Acer pseudo-platanus	30	3	3.4	4.7	2.6	*	*	*	0	100

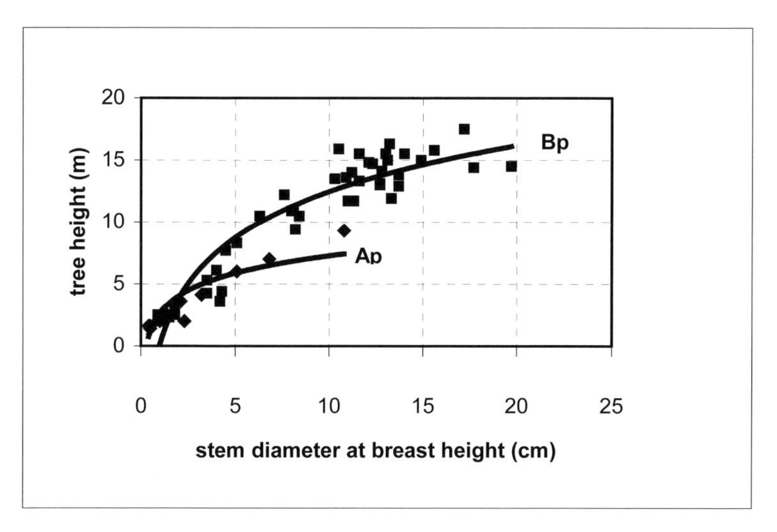

Fig. 4. Height versus diameter at breast height of *Betula pendula* (Bp) and *Acer pseudoplatanus* (Ap) in a 40-year-old birch stand at site WR

Ants and ground beetles

Do typical woodland fauna emerge in isolated urban-industrial woodlands? This question is addressed here using ants and ground beetles as indicator groups. The species composition in both groups shows significant differences between the pioneer, shrub and woodland sites (Figs. 6, 7). Generalists and species of dry, warm and open habitats dominate in the early stages. They are still present in the woodlands, but here, typical woodland species begin to dominate.

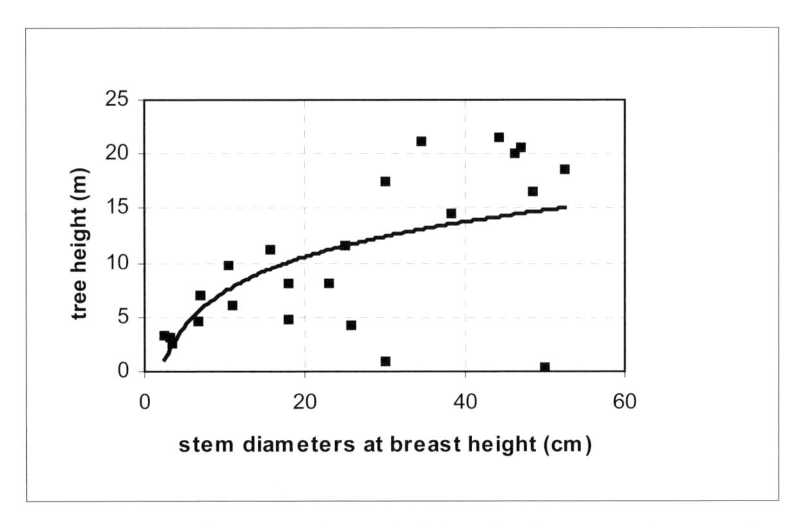

Fig. 5. Height versus diameter at breast height of *Robinia pseudoacacia* in an 80- to 90-year-old black locust stand at site WZ

There are obvious differences in the species richness of the two woodland sites. At the birch site WR, 14 ant species were found (12 with evidence of nests), at the black locust site WZ only 6 (5 with evidence of nests). Common species that are typical of the top layer of litter, such as *Stenamma debile*, are missing. Only three species of ground beetles were recorded at WZ over the three years of investigation. Two of these are typical woodland species that are capable of flying (see Table 4), the third is a xerophile species typical of open habitat conditions (*Harpalus rubripes*), and represents by far the most individuals. At the birch site WR, 11 ground beetle species have been recorded. Seven of these prefer woodlands, and they make up the majority of the individuals.

The differences between the two woodland sites may be due to the fact that site WZ is more isolated than site WR from old parks and remnants of the traditional open landscape. At WR, two of the woodland species are unable to fly (*Carabus nemoralis, C. coriaceus*), which results in reduced opportunities to colonise isolated habitats. In contrast, the two woodland species found at WZ were small and able to fly (Table 4). For ants, however, isolation is of less importance, because sexually mature individuals can fly. The low species richness at site WZ may be due to the low occurrence of greenflies (aphids) in the *Robinia* stands. Here, this important food resource of ants is confined to *Sambucus nigra* and *Acer pseudoplatanus*.

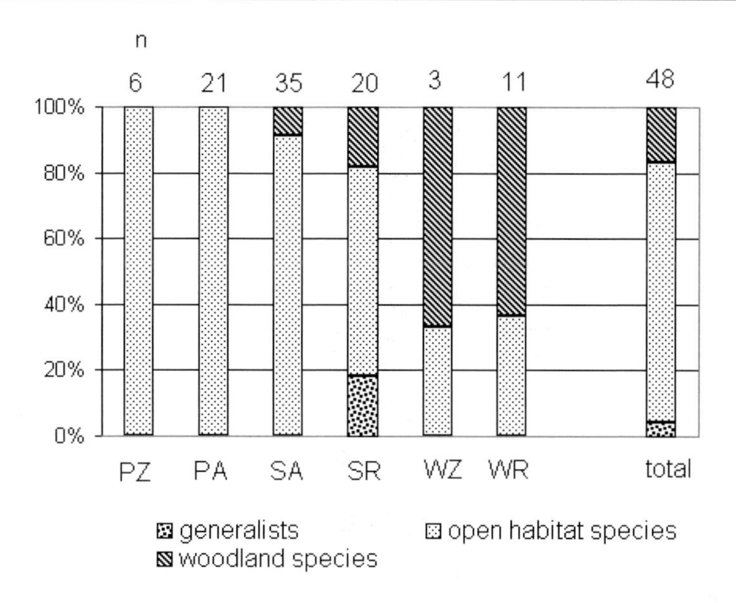

Fig. 6. Proportion of different ecological groups of ant species in pioneer, shrub and woodland plots (for abbreviations see Table 1). The results are the summary of data for 2000, 2001 and 2003; *n* indicates the total number of species

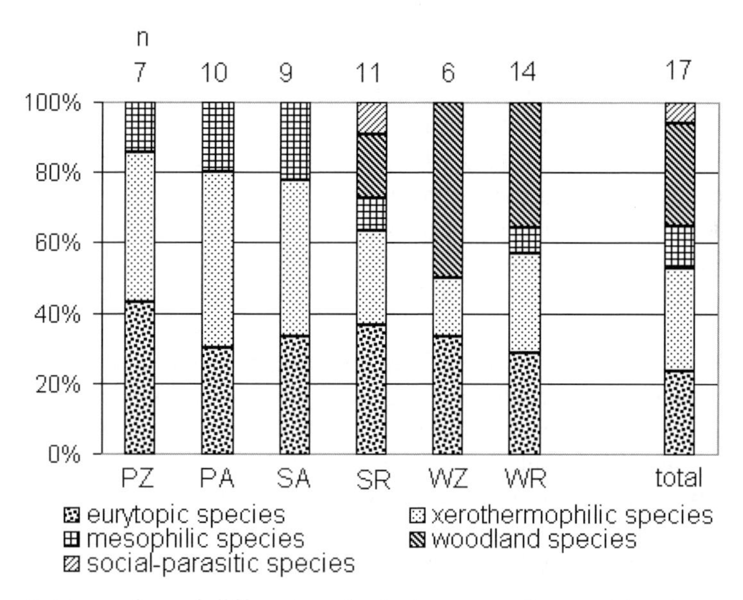

Fig. 7. Proportion of different ecological groups of ground beetle species in pioneer, shrub and woodland plots (for abbreviations see Table 1). The results are the summary of data for 2000, 2001 and 2003; *n* indicates the total number of species

Conclusions

Hard coal-mining spoil as soil parent material leads to extreme habitat conditions. The soils are very gravelly, stony and poor in fine earth and nutrients. Except on sites with fresh depositions of spoil, pH values are very low, which frequently increases the availability of heavy metals. This leads to slow migration of soil fauna and to slow decomposition of litter and soil development. Under these conditions, new types of plant communities arise on the post-industrial sites. The pioneer communities are usually rich in species that originate from a broad array of habitats (ruderal and salty habitats, river banks, arable fields, grasslands) or that were introduced from other regions. The comparison of different-aged sites shows that species richness will decline during succession. This also holds true for endangered early-successional species (Rebele and Dettmar 1996, Weiss and Schütz 1997, Weiss 2003b)

On the other hand, woodland species, both plants and animals, begin to colonise young woodland sites that are still dominated by pioneer trees such as birch and black locust. Examples include tree species (*Quercus robur, Acer pseudoplatanus, Carpinus betulus*), fern species (*Athyrium filix-femina, Dryopteris filix-mas, D. carthusiana*), and ground beetles (*Carabus nemoralis, C. coriaceus*). Differences between the plots may be attributed to isolation effects and to different food resources in stands of native (birch) versus non-native (black locust) tree species. In both woodland plots, early- and late-successional plant and animal species occurred together. This may be due to the immaturity of the woodlands and to their close connection to open habitats which is characteristic of urban-industrial woodlands. Similar results have been found when comparing the different succession stages of a derelict railway area in Berlin (Platen and Kowarik 1995). Further research will reveal whether the co-existence of different species groups remains a feature of the studied woodland communities.

A second open question is that of the direction of woodland succession. Late-successional tree species such as *Acer pseudoplatanus* or *Quercus robur* are already established under the canopy of birch- and black locust-dominated stands. It is doubtful, however, that these species will quickly outcompete the pioneer tree species. Obviously, the extreme soil conditions of the hard coal-mining spoils will remain unchanged for a long time. This may reduce the growth and competitiveness of species such as *Acer pseudoplatanus*. In the black locust stand, some of the taller trees have begun to die. This enhances the establishment of other tree species, but vege-

Table 4. Occurrence of ground beetles (carabids) on permanent plots in pioneer, shrub and woodland sites (0.1 ha; for abbreviations see Table 1). Data are the results of sampling by ground traps in 2000, 2001 and 2003. Column 2 (*W*, type of wings) gives information on species' dispersion ability: *b* brachypterous (unable to fly), *m* macropterous (able to fly), *d* dimorphous (both types occur)

	W	Pioneer sites						Shrub sites						Woodland sites					
		PZ			PA			SA			SR			WZ			WR		
		00	01	03	00	01	03	00	01	03	00	01	03	00	01	03	00	01	03
Total species richness		6	4	1	7	11	15	23	23	14	12	10	5	3	3	3	6	6	8
Total number of individuals		52	7	10	59	39	112	148	132	100	39	52	22	247	149	162	77	43	61
Woodland species richness		0	0	0	0	0	0	0	2	2	6	5	3	2	2	2	3	4	6
Total number of individuals		0	0	0	0	0	0	0	2	2	30	46	20	52	26	20	73	41	59
Carabus coriaceus	b	-	-	-	-	-	-	-	1	1	1	1	-	-	-	-	-	-	2
Carabus nemoralis	b	-	-	-	-	-	-	-	-	1	-	1	7	-	-	-	35	21	40
Leistus rufomarginatus	m	-	-	-	-	-	-	-	-	-	-	-	-	-	-	-	-	-	1
Nebria brevicollis	m	-	-	-	-	-	-	-	1	-	2	1	4	-	-	-	-	-	1
Notiophilus rufipes	d	-	-	-	-	-	-	-	-	-	2	3	-	4	3	14	13	5	2
Notiophilus biguttatus	d	-	-	-	-	-	-	-	-	-	23	40	9	48	23	6	25	14	13
Pterostichus strenuus	d	-	-	-	-	-	-	-	-	-	1	-	-	-	-	-	-	-	-
Pterostichus oblongopunctatus	m	-	-	-	-	-	-	-	-	-	1	-	-	-	-	-	-	1	-

tative regeneration of black locust also occurs in such stands and may prolong the predominance of the North American species (Kowarik 1996).

In addition to natural processes, the undirected succession of industrial abandoned land will be differentiated by the recreational activities of local people who make use of these new types of woodlands (Keil 2005). The quality of these sites is mostly due of the diverging patterns of pioneer, shrub and woodland stages and to the remnants of the industrial history. Thus, concepts for developing this post-industrial landscape should combine approaches that both enhance wilderness and maintain open habitats and cultural remnants. This is the main idea of the Projekt Industriewald Ruhrgebiet, which aims to enhance the social and ecological functions of the urban-industrial woodlands (Weiss 2003b, Dettmar 2005).

Acknowledgements

We would like to thank Dirk Hinterlang for the translation and Kelaine Vargas for the language editing. For the critical comments and valuable suggestions we would like to thank Ingo Kowarik.

References

AG Boden (1994) Bodenkundliche Kartieranleitung, 4th edn. E. Schweitzerbart'sche Verlagsbuchhandlung, Stuttgart

AK Stadtböden (1997) Empfehlungen des Arbeitskreises Stadtböden der Deutschen Bodenkundlichen Gesellschaft für die bodenkundliche Kartierung urban, gewerblich, industriell und montan überformter Flächen (Stadtböden). 2nd edn, Teil 1: Feldführer. Büro für Bodenbewertung, Kiel

Alef K (1991) Methodenhandbuch Bodenmikrobiologie – Aktivitäten, Biomasse, Differenzierung. Ecomed, Landsberg/Lech

Bouché M B (1972) Lombricien de France. Écologie et Systématique. Institut National de la Recherche Agronomique, Paris

Burghardt W (1989) C-, N- und S-Gehalte als Merkmale der Bodenbildung auf Bergehalden. Mitt Dtsch Bodenkundl Gesellsch 59:851–856

Burghardt W, Hiller D A, Stempelmann I, Tüselmann J (2003) Industriewald Ruhrgebiet – Bodenkundliche Untersuchungen. IÖR-Schriften 39:149–158

Dettmar J (2005) Forests for Shrinking Cities? The Project "Industrial Forests of the Ruhr". In: Kowarik I, Körner S (eds) Urban Wild Woodlands. Springer, Berlin Heidelberg, pp 263–276

Dunger W, Fiedler H J (1989) Methoden der Bodenbiologie. Gustav Fischer, Stuttgart

Fischer A (1987) Untersuchungen zur Populationsdynamik am Beginn von Sekundärsukzessionen. Diss Bot 110

Gausmann P, Keil P, Loos G, Haeupler H (2004) Einige bemerkenswerte floristische Funde auf Industriebrachen des mittleren Ruhrgebietes. Natur und Heimat Heft 2:47–54

Haeupler H, Kert C, Schürmann M (2003) Industriewald Ruhrgebiet – Floristisch-vegetationskundliche Untersuchungen. IÖR-Schriften 39:159–167

Hoffmann G (1991) Methodenbuch, Band 1: Die Untersuchung von Böden. VDLUFA, Darmstadt

Keil A (2005) Use and Perception of Post-Industrial Urban Landscapes in the Ruhr Area. In: Kowarik I, Körner S (eds) Urban Wild Woodlands. Springer, Berlin Heidelberg, pp 117–130

Kowarik I (1996) Funktionen klonalen Wachstums von Bäumen bei der Brachflächen-Sukzession unter besonderer Beachtung von *Robinia pseudoacacia*. Verh Ges Ökol 26:173–181

Kowarik I (2005) Wild urban woodlands: Towards a conceptual framework. In: Kowarik I, Körner S (eds) Urban Wild Woodlands. Springer, Berlin Heidelberg, pp 1–32

Kraft G (1884) Beiträge zur Lehre von den Durchforstungen, Schlagstellungen und Lichtungshieben. Hannover

Leder B, Leonhardt A (2003) Industriewald Ruhrgebiet – Untersuchungen zur Waldstruktur. IÖR-Schriften 39:169–177

Nobis M (1998) Vegetationsentwicklung auf anthropogen gestörten Sand- und Kiesböden der Oberrheinebene: Prognosen zur längerfristigen Sukzession auf Grundlage einer kurzfristigen Untersuchung (space-for-time substitution). Mitt Angewandte Ökologie: www.bwplus.fzk.de/paoedisk98

Oberdorfer E (1990) Pflanzensoziologische Exkursionsflora, 6th edn. Ulmer, Stuttgart

Pickett STA (1989) Space-for-time substitution as an alternative to long-term studies. In: Likens GE (ed) Long-term studies in ecology. Springer, New York, pp 110–135

Platen R, Kowarik I (1995) Dynamik von Pflanzen-, Spinnen- und Laufkäfergemeinschaften bei der Sukzession von Trockenrasen zu Gehölzstandorten auf innerstädtischen Bahnanlagen in Berlin. Verh Ges f Ökologie 24:431–439

Purtauf T, Dauber J, Wolters V (2004) Carabid communities in the spatio-temporal mosaic of rural landscape. Landscape and Urban Planning 67:185-193

Rebele F, Dettmar J (1996) Industriebrachen. Ulmer, Stuttgart

Schinner F, Öhlinger R, Kandeler E, Margesin R (1995) Methods in Soil Biology. Springer, Berlin Heidelberg

Schlichtig E, Blume H P, Stahr K (1995) Bodenkundliches Praktikum. 2nd den, Blackwell, Berlin Wien

Schulte A, Hamann M (2003) Industriewald Ruhrgebiet – Faunistische Untersuchungen. IÖR-Schriften 39:179–187

Tara K, Zimmermann K (1997) Brachen im Ruhrgebiet. LÖBF-Mitt 3/1997:16–21

VDLUFA (1991) Die Untersuchung von Böden. Handbuch der Landwirtschaftlichen Versuchs- und Untersuchungsmethodik, Band 1. VDLUFA, Darmstadt
Weiss J, Schütz P (1997) Effizienzkontrollen im Rahmen der Entwicklung von Industriebrachen. LÖBF-Mitt. 3/1997:22–27
Weiss J (2003a) Industriewald Ruhrgebiet – Daueruntersuchungen zur Sukzession auf Industriebrachen. IÖR-Schriften 39:139–147
Weiss J (2003b) „Industriewald Ruhrgebiet" – Freiraumentwicklung durch Brachensukzession. LÖBF-Mitt. 1/2003:55–59

Appendix

Table A1. Diaspores in the diaspore bank of shrub site SA on industrial land (rubble). *W* Results of seed washing method, *E* results of seed emergence method, *asterisks* indicate species which are common in the above-ground vegetation

Year of investigation	2000				2001				2003			
Date of collection	14 Jul		2 Oct		9 May		10 Sept		22 Jul		20 Sept	
Species / method	W	E	W	E	W	E	W	E	W	E	W	E
Achillea millefolium agg.*	-	-	-	-	-	-	-	-	-	-	1	8
Agrostis sp.	-	-	-	95	-	-	-	-	-	-	-	-
*Agrostis stolonifera**	-	2	-	-	-	-	5	-	-	-	-	-
*Anagallis arvensis**	-	-	1	1	-	-	-	-	-	-	1	1
Arenaria serpyllifolia	-	2	-	-	-	-	-	-	-	2	-	-
*Artemisia vulgaris**	-	-	-	-	-	-	1	-	-	-	-	-
Atriplex prostrata	-	-	-	-	-	-	1	-	-	-	-	-
*Betula pendula**	-	-	13	-	165	-	107	2	8	2	89	23
*Buddleja davidii**	-	-	-	-	-	-	-	-	-	-	2	-
Calystegia sepium	-	-	-	-	-	-	-	2	-	-	-	-
Cardamine hirsuta	-	1	-	-	-	-	-	1	-	-	-	-
Centaurium erythraea	-	-	-	-	-	-	5	3	-	-	-	-
*Centaurium pulchellum**	-	-	-	-	-	-	43	-	-	-	2	-
*Cerastium holosteoides**	-	-	-	-	-	-	-	-	-	-	1	-
Chenopodium glaucum	-	-	-	-	-	-	-	1	-	-	-	-
Chenopodium rubrum	-	-	-	-	-	-	37	-	-	-	-	-
*Cirsium arvense**	-	-	-	-	-	-	1	-	-	-	-	-
Conyza canadensis	-	-	-	-	-	-	-	-	-	-	1	1
*Daucus carota**	-	-	20	-	-	2	1	4	-	-	-	4
Echium vulgare	-	-	-	-	-	-	-	-	-	2	-	3
Epilobium ciliatum	-	-	-	1	-	-	-	-	-	-	-	-
*Epilobium hirsutum**	-	-	-	-	-	-	1	-	-	-	-	-
*Epilobium parviflorum**	-	-	-	-	-	-	1	-	-	-	-	-

Table A1. (cont.)

Epilobium spec.	-	-	-	-	-	-	-	5	-	-	-	-
Epilobium tetragonum	-	-	-	-	-	-	1	-	-	-	-	-
*Eupatorium cannabinum**	-	-	-	-	-	-	2	-	-	-	-	-
*Festuca rubra agg.**	-	-	-	-	-	-	-	-	-	3	-	2
Fragaria vesca	-	-	-	-	-	-	2	-	-	-	-	-
Galinsoga ciliata	-	-	-	-	-	-	-	2	-	-	-	-
Geranium molle	-	-	-	-	-	-	-	-	-	2	-	1
Hieracium sp.	-	1	1	-	-	-	-	-	-	-	-	-
*Holcus lanatus**	-	1	3	-	-	-	26	6	-	-	-	-
*Hypericum perforatum**	-	4	-	1	-	8	25	10	-	2	5	7
Hypericum sp.	-	1	1	-	-	-	-	-	-	-	-	-
Inula conyza	-	-	-	-	-	-	-	-	-	-	1	-
*Juncus tenuis**	-	-	1	-	-	-	14	-	-	-	-	-
*Plantago major**	-	-	-	-	-	-	4	-	-	1	1	7
Poa annua	-	2	-	2	-	-	1	-	-	-	-	-
Poa pratensis	-	-	-	-	1	-	-	1	-	-	-	-
Polygonum aviculare	-	-	-	-	4	-	23	-	8	-	-	-
Potentilla intermedia	-	-	-	-	-	3	3	4	-	-	-	-
*Potentilla norvegica**	-	12	-	1	5	9	7	8	3	-	29	-
Potentilla sp.	-	-	-	-	-	-	10	-	-	-	-	-
*Prunella vulgaris**	-	-	-	-	-	-	-	4	-	-	-	1
*Ranunculus repens**	-	-	-	-	-	-	-	2	-	-	-	-
Sagina procumbens	-	10	-	-	1	19	100	130	-	10	-	43
Senecio inaequidens	-	-	-	-	1	-	-	-	-	-	-	-
*Solidago gigantea**	-	1	26	1	-	-	3	-	-	-	-	-
Spergula arvensis	-	-	-	-	-	-	-	-	-	2	-	1
Spergularia rubra	-	-	-	-	2	-	-	-	-	-	-	-
Stellaria media	-	1	-	-	-	-	-	-	-	-	-	-
Poaceae spec.	-	-	-	-	-	-	-	-	-	-	-	3
Symphytum officinale	-	1		-	1	-	-	-	-	-	12	-
Taraxacum officinale agg.*	-	1	-	-	1	-	-	-	-	-	-	-
Tripleurospermun perforatum	-	1	-	-	-	-	1	-	-	-	-	-
*Veronica arvensis**	-	1	-	-	-	-	-	-	-	3	-	-
Veronica serpyllifolia	-	-	-	-	-	1	1	-	-	-	-	-
Vulpia myuros	-	-	-	-	-	-	-	-	-	2	2	-
Total diaspores and seedlings	-	38	66	102	179	42	427	189	19	31	157	105
Total taxa	-	13	8	7	7	6	26	17	3	11	13	14

Table A2. Diaspores in the diaspore bank of shrub site SR on industrial land (coal-mining spoil). *W* Results of seed washing method, *E* results of seed emergence method, *asterisks* indicate species which are common in the above-ground vegetation

Year of investigation	2000				2001				2003			
Date of collection	14 Jul		2 Oct		9 May		10 Sept		22 Jul		20 Sept	
Species/ method	W	E	W	E	W	E	W	E	W	E	W	E
*Acer pseudoplatanus**	-	-	-	-	-	-	-	-	-	-	1	-
Anagallis arvensis	-	-	-	-	-	-	-	-	-	-	4	1
Arenaria serpyllifolia	-	-	-	-	-	-		-	-	1	-	-
Artemisia vulgaris	-	-	1	-	-	-	-	-	-	-	1	1
*Betula pendula**	-	-	1091	-	726	-	953	-	214	13	627	134
Capsella bursa- pastoris	-	-	-	-	-	-	-	-	5	-	6	-
Carex sp.	-	-	-	-	1	-	-	-	-	-	-	-
Chenopodium polyspermum	-	-	1	-	-	-	-	-	-	-	-	-
Epilobium parviflorum	-	-	-	-	-	-	1	-	-	-	-	-
*Fragaria vesca**	-	-	91	1	62	1	24	1	-	-	-	-
Crepis capillaris	-	-	-	-	-	-	-	-	-	-	1	-
Hypericum perforatum	-	-	-	-	-	-	2	-	-	-	-	-
Lolium perenne	-	-	-	-	-	-	-	-	-	-	-	4
Medicago lupulina	-	-	-	-	-	-	-	-	30	-	58	-
Plantago major	-	-	-	-	-	-	-	-	-	-	5	-
*Poa annua**	-	-	-	1	-	-	-	-	-	1	-	-
Poaceae	-	-	-	-	-	-	-	-	3	-	-	-
Polygonum aviculare	-	-	6	-	-	-	-	-	-	-	-	-
Potentilla norvegica	-	-	4	-	-	-	-	-	-	-	-	-
Robinia pseudoacacia	-	-	-	-	2	1	2	-	-	-	-	-
*Rubus fruticosus agg.**	-	-	2	-	-	-	-	-	5	-	1	-
Rubus sp.	-	-	-	-	6	-	6	-	-	-	-	-
Sambucus nigra	-	-	-	-	-	-	5	-	-	-	-	-
Senecio inaequidens	-	-	1	-	-	-	-	-	-	-	-	-
*Solidago gigantea**	-	-	-	1	-	-	-	-	-	-	-	-
Total diaspores and seedlings	-	-	1197	3	797	2	993	1	257	15	704	140
Total taxa	-	-	8	3	5	2	7	1	5	3	9	4

Spontaneous Development of Peri-Urban Woodlands in Lignite Mining Areas of Eastern Germany

Sabine Tischew, Antje Lorenz

Vegetation Science and Landscape Ecology, University of Applied Sciences Anhalt

Introduction

In the eastern German federal states of Saxony, Saxony-Anhalt and Brandenburg, open-cast mining has destroyed vast landscapes of near-natural floodplain ecosystems and forests as well as elements of cultural landscapes or has affected them by lowering of the ground water table. On the one hand, forests have been considerably extended in total area compared to the situation before the open-cast lignite mining by the afforestation of spoil dumps. In Saxony and Saxony-Anhalt, for example, the destroyed forests have been replaced by 163 %. However, afforestation with monocultures of non-native species and low structural diversity has led to a mostly reduced ecological value of these forests (Berkner 1998). On the other hand, some spoil dumps have, more-or-less by chance, been settled by spontaneous succession since the 1930s. The result of this natural colonization is astonishing: on these sites, often called "lunar landscapes," grasslands, heaths and fens as well as varied woodlands have developed. Stages of development, species composition and stand structure of these woodlands are very different.

Due to their accessibility for urban residents and their location in post-industrial landscapes, these woodlands can be classified as "peri-urban woodlands." Species composition and stand structure mainly result from natural processes, but the site factors have been changed by the impact of open-cast mining. Therefore, these woodlands represent typical examples of "nature of the fourth kind" (see Kowarik 2005).

Thirty-two active surface mines were shut down after the German reunification in 1990. An area of nearly 1,000 km^2 was included in a reclamation process (LMBV 2002). Since the beginning of this process, scientists

Kowarik I, Körner S (eds) Wild Urban Woodlands.

and conservationists have demanded a stronger integration of natural processes in the development of woodlands. Gradually, this demand has been accepted by the reclamation company. About 3,400 ha of reclamation area are currently developing spontaneously into woodlands. With the designation of further successional areas, an additional 3,500 ha will develop into woodlands over different periods of time.

This paper summarizes the results of our studies on spontaneous woodland development on more than a hundred sites of eastern German lignite surface mines. The aim was to determine opportunities for the integration of spontaneous colonization processes into former mining areas. The present summary of results will focus on the following:

- the analysis of migration processes and the impact of isolating effects
- the analysis of site conditions and chronological differentiation of pioneer woodlands
- the integration of woodland development into general successional processes of post-mining landscapes
- opportunities and perspectives for peri-urban woodland development in post-mining landscapes

Locations and site characteristics

The study was carried out in nearly all eastern German brown-coal districts in Saxony, Saxony-Anhalt and Brandenburg that show spontaneous succession at least in some parts (FLB 2003; Tischew et al. 2004). Forty-four former mines were included in the investigations. Large areas with spontaneous succession were divided into smaller sites with relatively homogeneous conditions and considered as sample size. The time since abandonment of the investigated mines ranges from 1 to 98 years.

Lignite has been mined in surface mines in eastern Germany since the beginning of the 20th century. Overburden layers 40–120 m thick above the lignite seams had to be excavated and dumped, forming spoil dumps. This process can be compared with the mass turnover of the last glacial epoch (Müller and Eissmann 1991). The groundwater level had to be lowered significantly. Overburden layers consist of Tertiary and Quarternary substrates that were often mixed during the dumping processes. Quarternary substrates are usually more suitable for plant colonization. They are composed of boulder clay, sand and loess. Tertiary substrates consisting of sand, clay and silt often contain fine coal and sulphur. Due to the oxidation of the sulphuric compounds (iron pyrite, marcasite), Tertiary substrates tend to acidify over a long period of time as well as to release high levels of phytotoxic Al^{3+}-ions (Pichtel et al. 1988). Tertiary substrates additionally display high bulk density and low macropore volumina. High levels of finely distributed coal-containing admixtures can also influence substrate qualities negatively by their hydrophobic characteristics.

In the investigated mining area, the topsoil, which contained humus, was often not separated when dumped. Dumped substrates, therefore, show mostly low biological activity and low levels of available plant nutrients at the beginning of their development. Dumps of the early mining phases (up until the 1920s and 1930s) are exceptional, because they were turned by hand and topsoil containing humus was selectively put on the spoil dumps as the uppermost layer.

The area investigated has a mean annual temperature of 8.0–9.5°C and an annual precipitation of 450–650 mm. Due to industrial mining, vast surface mines have been created where isolating effects relating to colonization processes must be expected. Remnants of natural woodland vegetation (oak–hornbeam forests, beech forests, birch–oak forests, pine–oak forests, floodplain forests) have only rarely remained around spoil dumps.

Investigation methods

Investigations were carried out on 104 sample sites for the classification and description of chronologically and spatially differentiated successional series. Abundance and species-cover percentage of vascular plant species in all vegetation layers were estimated (modified Braun-Blanquet scale, Wilmanns 1998). In order to assess stand structure, all trees of 1 cm or greater diameter were measured in a plot of 400–2,000 m^2. Further investigations were carried out with respect to selected site factors, e.g. soil acidity, soil texture, nutrient availability, cation-exchange capacity, coal content, humus and fly-ash coating (for detailed methods, see Tischew et al. 2004). The classification of seral successional stages in space and time was carried out using hierarchical and non-hierarchical cluster analysis as well as multivariate ordination methods. The following bioindicative parameters were included for every plot: (1) percentage of vegetation layers, (2) weighted percentage of vegetation units (e.g. dry psammophytic grasslands, mesophilic grasslands, woodlands), (3) stand density per hectare of pioneer, intermediate and climax tree species and (4) stand basal area of all trees per hectare.

Special aspects of colonization processes were analyzed exclusively in brown-coal districts north and south of Halle: five northern mining areas from the region around Bitterfeld and five southern mining areas from

Geiseltal and its surroundings. Complete species composition was identified on sample sites (each about 2 km^2) by several mappings from 1998–2002. The investigated sample sites were characterized by open-land and woodland stages (age: 2–55 years). About 1,100 relevés were included, which were conducted during several research projects. To determine the regional species pool, we used the dataset of the flora mapping of the states Saxony-Anhalt and Saxony, which was done in grid cells (5.5 x 5.5 = 30.25 km^2).

Statistical tests and cluster analysis were carried out using SPSS 10.0 and PCORD 4.0. Multivariate ordination methods were carried out using CANOCO 4.5 (ter Braak and Šmilauer 1998).

Colonization processes

Contrary to other studies which report delayed colonization processes in restoration (Bakker et al. 1996; Coulson et al. 2001; Verhagen et al. 2001), more than 50% of the species found in the surrounding area of 30 km^2 were able to colonize the investigated mining sites north and south of Halle (Table 1). Depending on the landscape structure of the surroundings, up to 40% of the species already present in the mining sites were found more than 3 km away (long-distance dispersal). This phenomenon can be observed mainly at hospitable mining sites within a species-poor surrounding landscape (examples "southern mining sites").

High migration rates from larger distances can be explained by extraordinary events like gales, thermally induced turbulences or zoochory. These events can be compared to the migration processes after the last ice age (Clark 1998). Even rare species can accumulate at the mining sites which act as large "seed traps" in the cultural landscape (Tischew and Kirmer 2003). The availability of large-scale competition-free space in the mining sites supports the establishment of these species. Most of the species that migrated from distances greater than 3 km represent species from open landscapes and only 25% are woodland species. Most of these woodland species are trees and shrubs, woodland herbs are rather rare. However, woodland herbs are suitable for assessing successional dynamics in woodlands (Wolf 1989). Further analysis, therefore, refers particularly to this.

Table 1. Migration processes at ten former mines in Saxony-Anhalt. Comparison with the regional species pool and analysis of long-distance dispersal. *A* No. of species on the mining sites (mean size 1.9 km^2), *B* No. of species in the surroundings within a radius of 3 km (mean size 30 km^2), *C* No. of species migrated to the mining sites (according to the species pool of 3 km radius), *D* No. of species migrated to the mining sites from distances of more than 3 km, *E* No. of woodland species migrated to the mining sites from distances of more than 3 km. (SD=standard deviation, Mann-Whitney *U*-Test: *ns* = not significant; ** $0.01 \geq p > 0.001$; *** $p \leq 0.001$)

	A	B	C	D	E
Northern mining sites (n = 5)	298.8	642.2	263.4	35.4	9.0
SD (+/-)	56.5	69.7	56	5.5	2.8
Southern mining sites (n = 5)	258.4	292.6	154.0	104.4	23.8
SD (+/-)	38.0	34.9	24.7	23.4	8.1
Mann-Whitney-U-Test	ns	***	**	ns	ns

The relationship of distance of diaspore sources to the number of herb species in dump woodlands was analyzed in detail on four spoil dumps of similar age and similar site conditions (Fig. 1). In the case of the site with nearby diaspore sources (A), more than twice as many species migrated into the site than in the sites with more distant diaspore sources (sites B – D). The average number of species with animal dispersal (epizoochory, endozoochory) and self-dispersal (autochory) is much higher for the nearby site, while the average number of wind-dispersed (anemochorous) woodland herbs is nearly the same.

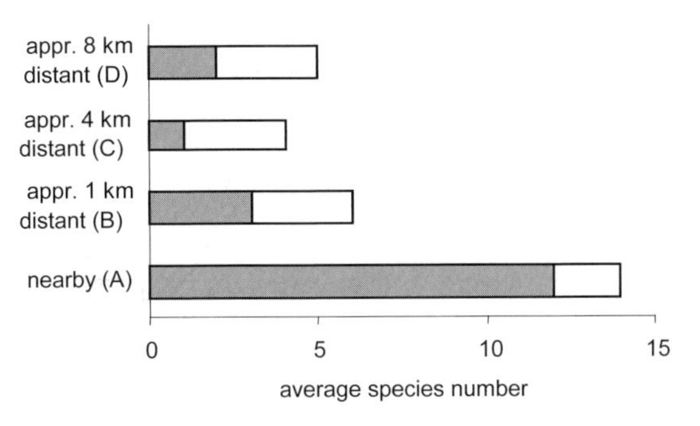

Fig. 1. Effect of the distance of diaspore sources on the species number of woodland herbs in pioneer woodlands on mining sites (plot size 400–525 m^2).

A Mining site Goitzsche-Tagesanlagen, *B* mining site Goitzsche Halde 1035, *C* mining site Roßbach, *D* mining site Kayna-Süd

More detailed investigations with respect to time-dependent colonization by woodland herbs were carried out based on data from all woodland areas investigated in eastern German post-mining landscapes. Table 2 shows a relatively high number of species that can already occur on young spoil dumps (10–60 years old). The seeds of these species were effectively dispersed by wind or animals over long distances (long-distance dispersal). Species with less effective dispersal mechanisms (autochory or dispersal by ants) mainly occur only on spoil dumps with pioneer woodlands, which are older than 60 years and characterized by an accumulation of intermediate and climax tree species in the shrub or tree layer (late-successional pioneer woodlands).

Indicator species of "ancient woodlands" (Peterken 1994; Wulf 1995) have a low frequency in general. They only occur in woodlands directly bordering "ancient woodlands" or on so-called "handmade dumps" (topsoil containing humus). It will probably take more time for isolated sites to be colonized by woodland species. Therefore, maintaining "ancient woodlands" in the surroundings of spoil dumps is an important basis for developing species-rich woodlands in post-mining landscapes (Benkwitz et al. 2002). In spontaneously developed woodlands, neophytes have rather a small share, with three species per plot. They comprise on average 2.8% of the herbaceous layer and 3.4% of the tree layer. Only the North American red oak (*Quercus rubra*) occurs more frequently in some regions. *Quercus rubra* was frequently planted in plantation forests, which provide a source for dispersal by jays.

Table 2. Frequency of selected woodland herbs in early and late successional pioneer woodlands with reference to their dispersal capacity and occurrence of indicator species for recent and ancient woodlands. **A**: number of occurrences (n_{total} = 104), **B**: frequency in pioneer woodlands < 60 years (%, n_{total} = 86), **C**: frequency in woodlands > 60 years (%, n_{total} = 18), **D**: dispersal modes (Müller-Schneider 1986; Bonn et al. 2000): a – anemochorous, z_{epi} – epizoochorous, z_{endo} – endozoochorous, m – myrmecochorous, s – autochorous, ? – dispersal mode unknown, **E**: Long-distance dispersal capacity (Frey and Lösch 1998; Verheyen et al. 2003), long – present; no – not present, **F**: indicator species for r: recent woodlands and a: ancient woodlands (Wulf 1995, 2003).

	A	B	C	D	E	F
Aegopodium podagraria	2	1.2 %	5.6 %	s	no	r
Anemone nemorosa	1	0 %	5.6 %	s, m	no	a
Avenella flexuosa	27	26.7 %	22.2 %	a, z_{epi}	long	r
Brachypodium sylvaticum	26	12.8 %	83.3 %	a, z_{epi}	long	a
Calamagrostis arundinacea	1	1.2 %	0 %	a, z_{epi}	long	
Campanula trachelium	2	0 %	11.1 %	s	no	a

Table 2. (cont.)

Convallaria majalis	4	1.2 %	16.7 %	s, z_{endo}	no & long	a
Carex brizoides	2	1.2 %	5.6 %	a, z_{epi}	long	
Carex montana	5	3.5 %	11.1 %	a, z_{epi}, m	no & long	
Carex pilulifera	10	10.5 %	5.6 %	a, z_{epi}, m	no	a
Circaea lutetiana	2	0 %	11.1 %	z_{epi}	long	a
Clematis vitalba	5	2.3 %	16.7 %	a, z_{epi}	long	
Dactylis polygama	6	5.8 %	5.6 %	a, z_{epi}	long	
Dryopteris carthusiana	3	2.3 %	5.6 %	a	long	a
Dryopteris filix-mas	4	2.3 %	11.1 %	a	long	
Epilobium montanum	5	4.7 %	5.6 %	a	long	
Epipactis atrorubens	22	18.6 %	33.3 %	a	long	
Festuca gigantea	4	0 %	22.2 %	a, z_{epi}	long	a
Festuca heterophylla	1	0 %	5.6 %	a, z_{epi}	long	
Fragaria vesca	37	29.1 %	66.7 %	z_{endo}	long	
Hedera helix	3	1,2 %	11.1 %	z_{endo}	long	
Hieracium lachenalii	54	55.8 %	33.3 %	a	long	a
Hieracium laevigatum	40	39.5 %	33.3 %	a	long	
Hieracium murorum	20	19.8 %	16.7 %	a	long	
Hieracium sabaudum	76	74.4 %	66.7 %	a	long	
Hieracium umbellatum	1	1.2 %	0 %	a	long	
Holcus mollis	1	1.2 %	0 %	a, z_{epi}	long	
Listera ovata	6	7.0 %	0 %	a	long	
Luzula pilosa	2	0 %	11.1 %	m	no	a
Maianthemum bifolium	3	0 %	16.7 %	z_{dyso}	no	a
Melampyrum pratense	9	8.1 %	11.1 %	m	no	a
Melica nutans	2	0 %	11.1 %	m	no	a
Milium effusum	1	0 %	5.6 %	m, z_{epi}	no & long	a
Moehringia trinervia	4	0 %	22.2 %	m	no	a
Monotropa hypophegea	3	2.3 %	5.6 %	?	?	
Orthilia secunda	16	12.8 %	27.8 %	a	long	
Poa nemoralis	27	25.6 %	27.8 %	a, z_{epi}	long	a
Polygonatum multiflorum	1	0 %	5.6 %	z_{epi}	no	a
Pyrola chlorantha	1	1.2 %	0 %	a	long	
Pyrola minor	18	15.1 %	27.8 %	a	long	
Pyrola rotundifolia	1	0 %	5.6 %	a	long	
Scrophularia nodosa	1	0 %	5.6 %	s	no	
Solidago virgaurea	10	10.5 %	5.6 %	a, z_{epi}	long	
Stachys sylvatica	1	0 %	5.6 %	z_{epi}	long	a
Sanicula europaea	2	0 %	11.1 %	z_{epi}	long	a
Stellaria holostea	2	0 %	11.1 %	z_{epi}	no?	a
Vacinium myrtillus	10	5.8 %	27.8 %	z_{endo}	no & long	a
Vacinium vitis-idea	7	3.5 %	22.2 %	z_{endo}	no & long	
Viola riviniana/V. reichenbachiana	15	7.0 %	50.0 %	s, m	no	a

Site-dependent and chronological woodland differentiation

Based on multivariate analysis, three successional series (A, B, C) could be derived for woodland development on different site conditions (Fig. 2). Only woodlands not directly influenced by groundwater were included in the analysis.

Woodland development proceeds very slowly on sites with a high proportion of Tertiary substrates (A). These are characterized by extremely acid pH-values (minimum 3,0), high to very high coal contents (25 % on average in the 60–100 years old stages) and hydrophobic features. The carbon-to-nitrogen ratio as one parameter for nutrient availability was extremely high, even on the older dumps (1:55 in a soil depth of 0–10 cm in the 60–100 years old stages). These substrates are hardly suitable for colonization. Several pioneer woodland stages have to be passed through frequently until the development to later woodland stages is possible.

Considerably faster development compared to species-rich woodlands was recorded on more hospitable substrates depending on diaspore sources in the surroundings (B, C). Sites of series B are characterized by Tertiary and Quarternary mixed substrates. They have low to moderate pH-values (3,9–5,9; all data relate to the upper soil layers (0–10 cm depth) in the 60–100 years old woodland stages) and low to moderate nutrient availability (C/N ratio 1:23 to 1:40). An initial accumulation of intermediate and climax tree species in the tree and shrub layer as well as first woodland herbs in the herbaceous layer can be observed in the age-class 60 - 100 years. The quickest successional progress was recorded in the woodlands of the mesophilic to nutrient rich sites of series C. The Tertiary and Quarternary mixed substrates show moderate to high pH-values (4.7 to 6.3) and moderate to high nutrient availability (C/N ratio 1:17 to 1:29). Compared to series B, an even greater accumulation of particular woodland herbs in the herbaceous layer as well as intermediate and climax tree species in the tree and shrub layer may be recorded after a 60–100 years development.

Fig. 2 (next page) Successional series (A, B, C) in lignite mining areas of eastern Germany

successional series	**series A** extreme sites	**series B** mesophilic sites	**series C** mesophilic to nutrient rich sites
characteristics	- tertiary sand or silt with extremely acid pH-value - high to very high coal contents, very low nutrient availability	- mixed substrates (tertiary or quarternary materials) - low (to moderate) pH-value low (to moderate) nutrient availability	- quarternary or mixed substrates - moderate (to high) pH-value, moderate (to high) nutrient availability
10-30 years	**A-1:** Initial openland-shrub stages with birch and/or pine 	**B-1:** Initial pioneer birch woodlands by skipping or quickly passing the stage of herbaceous layer 	**C-1:** Initial pioneer birch woodlands with simultaneous development of a herbaceous layer 
30-60 years	**A-2:** Very thin pioneer woodlands with birch and/or pine 	**B-2:** Pioneer birch woodlands without accumulation of intermediate/climax tree species 	**C-2:** Pioneer birch woodlands with initial accumulation of woodland herbs and intermediate/climax tree species in the shrub layer 
60-100 years	**A-3:** Persisting thin pioneer woodlands with birch and/or pine 	**B-3:** Pioneer birch woodlands with some woodland herbs and initial accumulation of intermediate/climax tree species in the shrub/tree layer 	**C-3:** Pioneer birch woodlands with accumulation of woodland herbs and intermediate/climax tree species in the shrub/tree layer 

In addition to site conditions, successional rate depends on availability of diaspore sources in the surroundings and substrate-modifying influences (e.g. fly-ash coating). Non-linear developments, related to general individual conditions are possible, i.e. successional stages can be passed through more quickly or slowly.

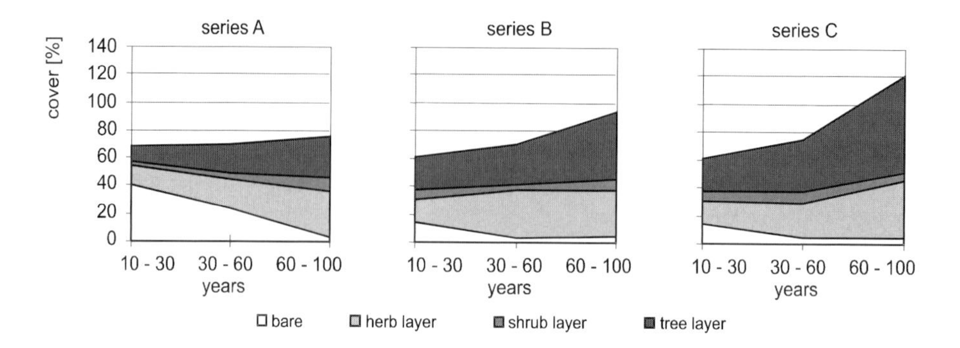

Fig. 3. Development of stand structure (bare soil, herbaceous, shrub and tree layers) of the successional series A, B and C

As shown in Fig. 3, bare soils and open woodland structures persist on the less hospitable sites of series A over a considerably longer period, in contrast to series B and C. This development is not necessarily negative with respect to biodiversity as these woodlands remain as refuges for the less competitive, rare species which prefer open woodland structures. This differentiated development also leads to a varied and aesthetically attractive landscape.

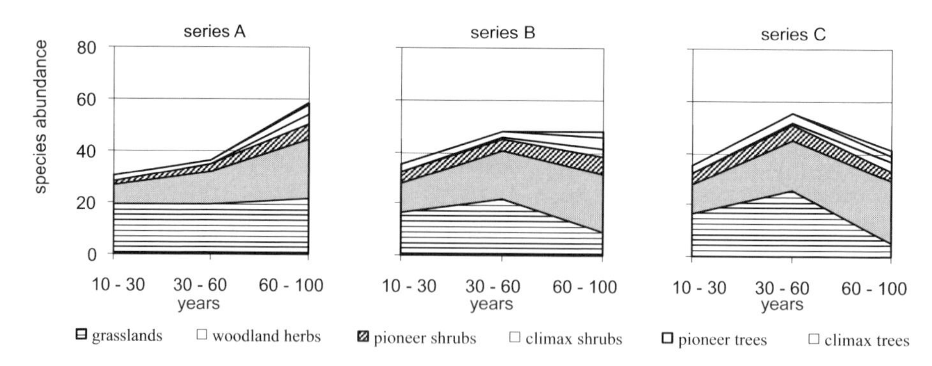

Fig. 4. Development of species number in herbaceous, shrub and tree layers of the successional series A, B and C

Fig. 4 presents differences in the development of species diversity in separated vegetation layers. High numbers of woodland species in shrub and tree layers, as well as herbaceous layers, mainly occur on the oldest sites of series C. Many of these spoil dumps represent "handmade dumps." Here, woodland species can establish themselves faster on account of the topsoil coverings, which accelerate soil formative processes and provide a diaspore source of the woodland species. Therefore, slower development processes are expected on isolated younger (non-handmade) sites of this successional series in the future. However, the younger and less developed age-classes of succession are also of high value for nature conservation and for experiencing nature. For example, a total of 245 species are found on a 20 ha, 30-year-old dump with different stages of birch and pine pioneer woodlands on the Goitzsche mining site. Fourteen species alone are to be found on Red Lists of Saxony-Anhalt (Frank et al. 1992) and Germany (Korneck et al. 1996). Therefore, young woodland stages in post-mining landscapes contribute substantially to the maintenance of biological diversity in post-industrial landscapes.

Gradual migration of woody species onto spoil dumps was evaluated in detail for successional series C (Fig. 5). Young stages are characterized by the pioneer tree-species silver birch (*Betula pendula*) and Scots pine (*Pinus sylvestris*). Common oak (*Quercus robur*) can migrate onto young sites, whereas most intermediate and climax tree species become established only slowly in the shrub and tree layer in the second or third stage. Rejuvenation of pioneer tree species is already considerably restricted on the oldest sites. Here, a general change in stand structure is beginning as pioneer tree species are also dying off.

Woodland development during succession on spoil dumps – an overview

Further data (FLB 2003; Tischew et al. 2004) were included to integrate woodland development into general colonization processes on soil dumps. Ordination methods or results from permanent plots were used to determine site factors (e.g. pH values, soil textures, diaspore sources) that could work as "switch points" within successional dynamics (Fig. 6). Based on these results and on the development of complex successional networks, it is possible to predict general developmental tendencies in eastern German post-mining landscapes. Depending on the diaspore sources in the direct

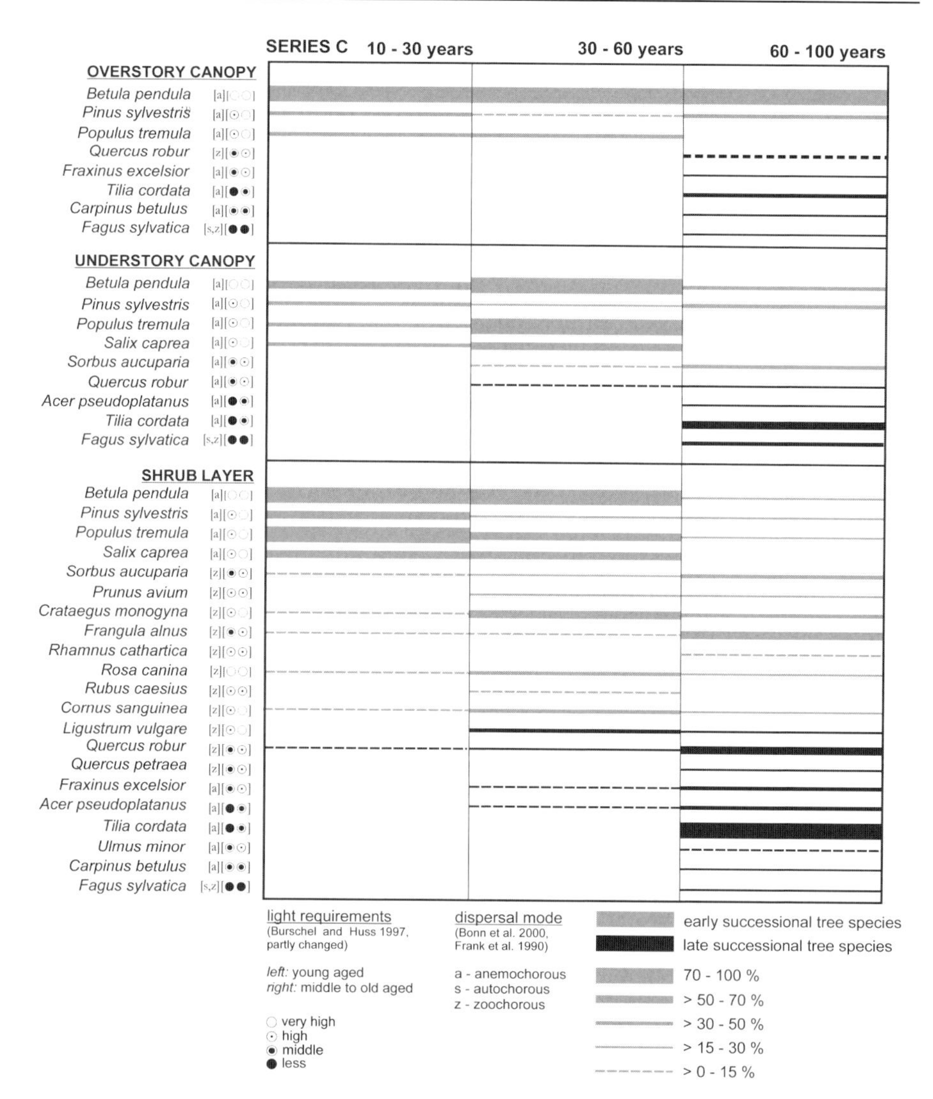

Fig. 5. Frequency of selected pioneer, intermediate and climax tree species in shrub as well as upper and lower tree layers in three age-classes (10–30 years, 30–60 years, 60–100 years) of the successional series C, with information about dispersal modes and light requirements in different life stages

Fig. 6 (next page) Successional scheme for primary succession in eastern German post-lignite mining landscapes on sites distant from groundwater (groundwater depth > 2 m)

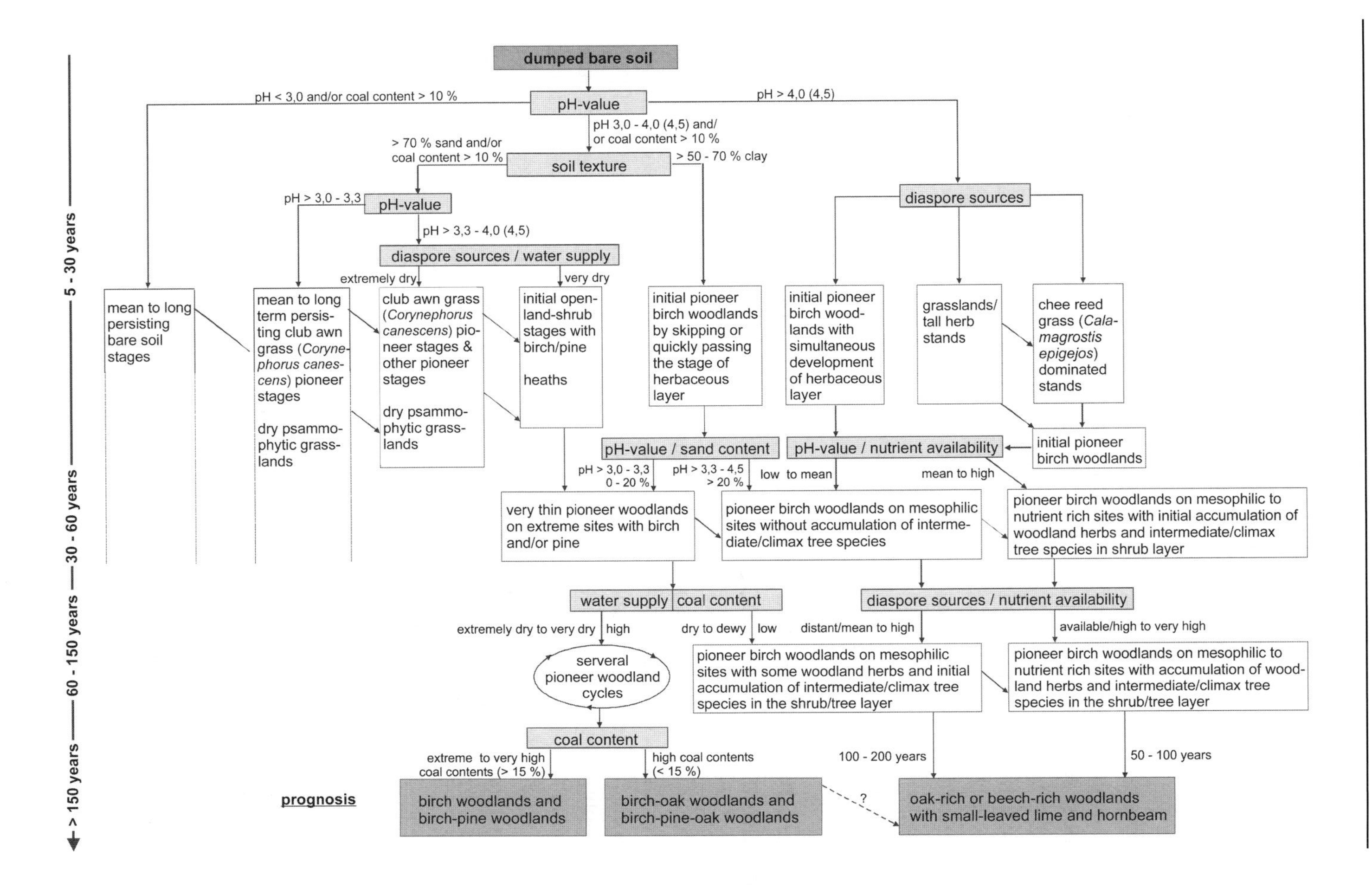

dumped bare soil
pH-value
pH < 3,0 and/or coal content > 10 %
pH 3,0 - 4,0 (4,5) and/ or coal content > 10 %
pH > 4,0 (4,5)
> 70 % sand and/or coal content > 10 %
soil texture
> 50 - 70 % clay
pH > 3,0 - 3,3
pH-value
pH > 3,3 - 4,0 (4,5)
diaspore sources / water supply
extremely dry
very dry
diaspore sources
mean to long persisting bare soil stages
mean to long term persisting club awn grass (Corynephorus canescens) pioneer stages
dry psammophytic grasslands
club awn grass (Corynephorus canescens) pioneer stages & other pioneer stages
dry psammophytic grasslands
initial open-land-shrub stages with birch/pine
heaths
initial pioneer birch woodlands by skipping or quickly passing the stage of herbaceous layer
initial pioneer birch woodlands with simultaneous development of herbaceous layer
grasslands/ tall herb stands
chee reed grass (Calamagrostis epigejos) dominated stands
initial pioneer birch woodlands
pH-value / sand content
pH > 3,0 - 3,3 0 - 20 %
pH > 3,3 - 4,5 > 20 %
pH-value / nutrient availability
low to mean
mean to high
very thin pioneer woodlands on extreme sites with birch and/or pine
pioneer birch woodlands on mesophilic sites without accumulation of intermediate/climax tree species
pioneer birch woodlands on mesophilic to nutrient rich sites with initial accumulation of woodland herbs and intermediate/climax tree species in shrub layer
water supply
coal content
diaspore sources / nutrient availability
extremely dry to very dry
high
dry to dewy
low
distant/mean to high
available/high to very high
serveral pioneer woodland cycles
pioneer birch woodlands on mesophilic sites with some woodland herbs and initial accumulation of intermediate/climax tree species in the shrub/tree layer
pioneer birch woodlands on mesophilic to nutrient rich sites with accumulation of woodland herbs and intermediate/climax tree species in the shrub/tree layer
coal content
extreme to very high coal contents (> 15 %)
high coal contents (< 15 %)
100 - 200 years
50 - 100 years
prognosis
birch woodlands and birch-pine woodlands
birch-oak woodlands and birch-pine-oak woodlands
?
oak-rich or beech-rich woodlands with small-leaved lime and hornbeam
5 - 30 years
30 - 60 years
60 - 150 years
> 150 years

surroundings, species that arrive first on hospitable sites determine the vegetation structures on these sites (right-hand side of the diagram). Initial colonization patterns are also influenced by the availability of so-called safe sites (e.g. small depressions accumulating water and a little organic material). These initial patterns lead to secondary patterns by the mechanisms of "facilitation" or "inhibition" among the already established species and the newly immigrated species, and can vary a lot (patch-dynamics concept, Wiegleb and Felinks 2001; Prach 1987). Differentiation processes of species based on site factors become clear in older successional stages. On extreme sites, the influence of stochastic effects is, in principle, smaller due to selection for species that can tolerate these site factors (e.g. extremely low pH values, very wet or very dry soils). Vegetation development is impeded mainly on dry and extremely acidic sites (left-hand side of the diagram), where successional stages free of trees are expected over longer periods.

Opportunities and perspectives for woodland development in post-mining landscapes

As a result of the closure of many mines at the beginning of the 1990s after the German Reunification, there are still large areas of bare soil dumps as well as older open-cast areas. Based on the analysis of site conditions, spontaneous development of large woodlands is to be predicted for many parts of surface mines. These woodlands will show high chronological and spatial differentiation in their development due to site heterogeneity and different site advantages. This is related to different land-use potentials. Varied and continuously changing landscape structures offer ideal conditions for recreational activities in the sense of "nature tourism." The mainly regional population uses this wilderness for walking, cycling or mushrooming. The unique potential of these sites in terms of process conservation must also be maintained for nature conservation. In general, greater proportions of abandoned lignite mining areas reserved for spontaneous succession will increase the floristic diversity in intensively used landscapes (Hadacova and Prach 2003; Tischew and Kirmer 2003). Conservation organizations and foundations are taking this chance and currently buying up large successional areas. Conservation authorities in the federal states support them. As scientists see it, these sites can be used as unique specimens for the observation of regeneration processes in large, disturbed areas. Spontaneous woodland development on these sites should also be encouraged for financial reasons: afforestation costs about

€ 20,000/ha and mostly can not be sold or cultivated profitably, whereas succession sites require only about € 1,400/ha (Abresch et al. 2000). Suitable methods for near-natural acceleration of vegetation development have been developed for problem areas, where vegetation development has to be accelerated because of erosion or to avoid dust emissions (Tischew et al. 2004).

Summary

We studied the spontaneous development of peri-urban woodlands on more than a hundred sample sites in eastern German open-cast lignite mining areas. The regeneration ability of the sites and the variety of developmental paths are remarkable. On hospitable substrates, pioneer birch forests can spontaneously develop within a few years if diaspore sources are available. Intermediate and climax tree species (e.g. common oak, small-leaved lime, sycamore) take root on these substrates on 30-year-old sites. They slowly displace birch after 60–80 years. Due to the variety of substrates and topography, spoil-dump woodlands are mostly of high structural diversity and aesthetically enrich post-mining landscapes. Thin birch and pine woodlands, including grasslands or heaths on less hospitable sites, are interesting examples of colonization processes after a massive turnover which was previously found to this extent only during the last ice age. Less competitive species can find refuges on these sites for longer periods. Therefore, woodlands in post-mining landscapes contribute to the maintenance of biological diversity and dynamic processes in post-industrial landscapes. Though far-reaching disturbance is the starting point of its development, a unique "nature of the fourth kind" can develop by natural colonization processes and offer various options for recreational activities for the public to experience a natural wilderness as well as potential for nature conservation.

Economical points of view are also good reasons for including spontaneous colonization processes in the design of former surface-mining areas. Spontaneous succession in surface mines therefore represents a great challenge and requires a rethinking of reclamation strategies by all people and institutions involved. More than 10,000 ha of eastern German post-mining landscapes are currently designated as sites for succession.

Acknowledgements

Investigations were funded by the German Federal Ministry for Education, Science, Research and Technology (BMBF) as well as the Lusatian and Central German Mining Management Company (LMBV).

References

Abresch JP, Gassner E, von Korff (2000) Naturschutz und Braunkohlensanierung. Angewandte Landschaftsökologie 27:427ff

Bakker JP, Poschlod P, Stryskstra RJ, Bekker RM, Thompson K (1996) Seed banks and seed dispersal: important topics in restoration ecology. Acta Botanica Neerlandica 45 (4):461–490

Benkwitz S, Tischew S, Lebender A (2002) „Arche Noah" für Pflanzen? Zur Bedeutung von Altwaldresten für die Wiederbesiedlungsprozesse im Tagebaugebiet Goitzsche. Hercynia N.F. 35:181–214

Berkner A (1998) Das Mitteldeutsche Braunkohlenrevier - Naturraum und ausgewählte Geofaktoren im Mitteldeutschen Förderraum – Ausgangszustand, bergbaubedingte Veränderungen, Zielvorstellungen. – In: Pflug W (ed) Braunkohlentagebau und Rekultivierung: Landschaftsökologie, Folgenutzung, Naturschutz. Springer, Berli Heidelberg, pp 767–779

Bonn S, Poschlod P, Tackenberg O (2000) Diasporus – a database for diaspore dispersal – concept and applications in case studies for risk assessment. Ökologie und Naturschutz 9:85 –97

Clark J (1998) Why trees migrate so fast: confronting theory with dispersal biology and the paleorecord. Amer Nat 152:204–224

Coulson SJ, Bullock JM, Stevenson MJ, Pywell RF (2001) Colonization of grassland by sown species: dispersal versus microsite limitation in responses to management. J Appl Ecol 38:204–216

FLB (2003) Forschungsverbund Landschaftsentwicklung Mitteldeutsches Braunkohlenrevier: Analyse, Bewertung und Prognose der Landschaftsentwicklung in Tagebauregionen des Mitteldeutschen Braunkohlenrevieres. Final report University of Applied Sciences, Anhalt

Frank D, Herdam H, Jage H, Klotz S, Rattey S, Wegener U, Weinert U, Westhus W (1992) Rote Liste der Farn- und Blütenpflanzen des Landes Sachsen-Anhalt. Unter Mitwirkung von T Benkert, S Bräutigam, H Kahlmeyer, H-U Kisson, J Peterson, J Pusch, G Stohr. Berichte des Landesamtes für Umweltschutz Sachsen-Anhalt 1:44–63

Frey W, Lösch R (1998) Lehrbuch der Geobotanik – Pflanze und Vegetation in Raum und Zeit. Gustav Fischer, Ulm

Hadacova D, Prach K (2003) Spoil heaps from brown coal mining: Technical reclamation versus spontaneous revegetation. Restoration Ecology 11(3):385–391

Korneck D, Schnittler M, Vollmer I (1996) Rote Liste der Farn- und Blütenpflanzen (Pteridophyta et Spermatophyta) Deutschlands. Schriftenreihe für Vegetationskunde 29:21–187

Kowarik I (2005) Wild urban woodlands: Towards a conceptual framework. In: Kowarik I, Körner S (eds) Urban Wild Woodlands. Springer, Berlin Heidelberg, pp 1–32

LMBV (2002) Sanierungsbericht 2001. LBMV Presse und Öffentlichkeitsarbeit Berlin

Müller A, Eissmann L (1991) Die geologischen Bedingungen der Bergbaufolgelandschaft im Raum Leipzig. In: Hänsel C (ed) Umweltgestaltung in der Bergbaufolgelandschaft. Sonderband Abhandl. Sächs. Akad. d. Wiss. zu Leipzig. Akademie-Verlag, Berlin, pp 39-45

Müller-Schneider P (1986) Verbreitungsbiologie der Blütenpflanzen Graubündens. Veröff Geobot Inst ETH Stiftung Rübel 85:5–263

Peterken GF (1994) The definition, evaluation and management of ancient woodlands in Great Britain. NNA-Ber 3:102–114

Pichtel JR, Dick WA, Hanna RM (1988) Reclamation of a marcasite soil by addition of limestone. Bull New Jersey Acad Sci 33:7–14

Prach P (1987) Succession of vegetation on dumps from strip coal mining, N.W. Bohemia, Czecheslovakia. Folia Geobotanica et Phytotaxonomica 22:339–354

ter Braak CJF, Šmilauer P (1998) CANOCO reference manual and user's guide to Canoco for Windows. Software for canonical community ordination (version 4). Centre for Biometry, Wageningen

Tischew S, Kirmer A (2003) Entwicklung der Biodiversität in Tagebaufolgelandschaften: Spontane und initiierte Besiedlungsprozesse. Nova Acta Leopold NF 87 (328):249–286

Tischew S, Lorenz A, Striese G, Benker J (2004) Analyse, Prognose und Lenkung der Waldentwicklung auf Sukzessionsflächen der Mitteldeutschen und Lausitzer Braunkohlenreviere. Unpublished research report, University of Applied Sciences, Anhalt

Verhagen R, Klooker J, Bakker JP, van Diggelen R (2001) Restoration success of low-production plant communities on former agricultural soils after top-soil removal. J Appl Veg Sci 4:75–82

Verheyen K, Bossuyt O, Hermy M (2003) Herbaceous plant community structure of ancient and recent forests in two contrasting forest types. Basic Appl Ecol 4:537–546

Wiegleb G, Felinks B (2001) Primary succession in post-mining landscapes of Lower Lusatia – chance or necessity. Ecological Engineering 17:199–217

Wilmanns O (1998) Ökologische Pflanzensoziologie, 6. Auflage. UTB-Verlag Quelle u. Meyer, Heidelberg

Wolf G (1989) Probleme der Vegetationsentwicklung auf forstlichen Rekultivierungsflächen im Rheinischen Braunkohlenrevier. Natur und Landschaft 64(10):451–455

Wulf M (1995) Historisch alte Wälder als Orientierungshilfe zur Waldvermehrung. LÖBF-Mittlg. 4/1995:62–71

Wulf M (2003) Preference of plant species for woodlands with differing habitat continuities. Flora 198:444–460

Ecological Networks for Bird Species in the Wintering Season Based on Urban Woodlands

Tomohiro Ichinose

Institute of Natural and Environmental Science, University of Hyogo

Background

We can watch many birds in urban woodlands. Recently urban woodlands have been considered an important habitat for bird species, and many studies have been conducted in order to identify which environmental factors of urban woodlands and open space are necessary for bird habitats, for example, size, shape, vegetation type and so on. Many studies have indicated that woodland size is the most important factor (e.g., Martin 1983), however it is always very difficult to establish a new large open space in urban areas, especially in Japan, because of the high price of land. In recent decades, the ecological network concept has been taken into account in urban and rural planning (e.g., Haase et al. 1992); it aims to provide the physical conditions that are necessary for populations of species to survive within a landscape that, to a greater or lesser extent, is also exploited by economic activities (Nowicki et al. 1996). If the ecological network between woodlands is improved and conserved in urban areas, more bird species can live and breed there.

We have no fixed method in urban areas for planning for ecological networks. Kirby (1994) indicated that few field studies could show the effectiveness of ecological networks for animals and plants. Recently in Japan, some ecological network plans have been created in urban areas; however, the environmental factors that are specifically important for certain taxa or species has seldom been considered during planning.

The objective of this study is to identify relationships between the presence of bird species and environmental factors in urban areas for ecological network planning. Vegetation cover of urban parks is addressed in depth.

Kowarik I, Körner S (eds) Wild Urban Woodlands.

Material and methods

Study area

The study area was the southern part of Nishinomiya City, Hyogo Prefecture, central Japan (Fig. 1), located between Osaka and Kobe. Commercial and residential land uses are dominant in the area, though there are many urban woodlands consisting of fragmented forests, shrine and temple forests, riparian forests, urban parks and so on. Twenty urban parks with vegetation cover were selected as study sites. The area of the smallest park was 0.19 ha while that of the largest one was 3.4 ha (the median, and first and third quartiles were 0.55, 0.305, and 1.09 ha, respectively).

Bird censuses

Birds in each park were censused from 13 December 1999 to 23 February 2000. This period was the wintering season in this study area. I censused birds three times in each park. I used the point-counts method described by Bibby et al. (1992). I recorded all species observed within a 25 m radius from the standing point and omitted the individuals flying beyond the observation range. Birds were censused for 20 minutes between 0900 and 1500, and the species name, the number of individuals, and the observed position (height from ground and plant species) were recorded.

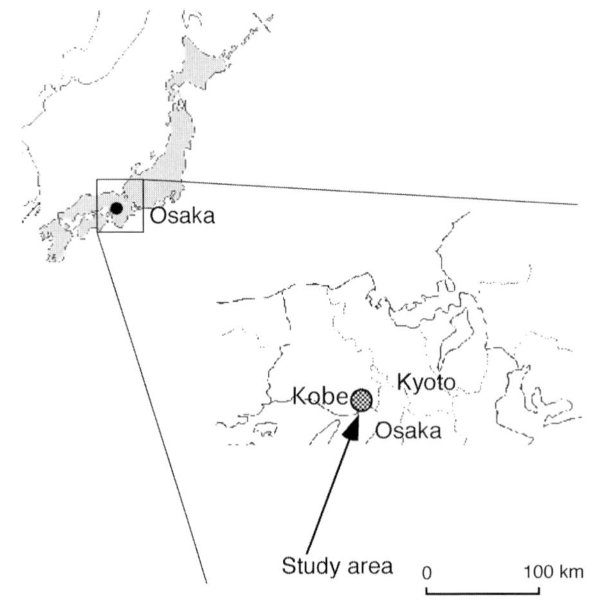

Fig. 1. Location of the study area

Vegetation structure measurements

In each park, a 50 m transect was established, located essentially at random but covering all representative structures. Vegetation structure and composition were recorded at 25 points (spaced at 2 m intervals) along the line. The presence or absence of vegetation within a circle of 0.60 m diameter was recorded at each point, at thirteen different heights: 0, 0.12, 0.37, 0.75, 1.5, 3.0, 4.5, 6, 9, 12, 18, 24 and > 32 m (cf., Erdelen 1984). In this way, 25 recordings (presence or absence) were obtained for each of the 13 heights per urban park. In all recordings with vegetation present, the respective plant species were also determined.

The following variables were defined for each park:

- PB00 to PB12: plant-body amount = for each of the 13 layers, the number of positive recordings (i.e., vegetation present) along the transect (maximum = 25 points per layer). The term "layer" does not imply the presence of a distinct and recognizable structural element of vegetation, but simply refers to the levels at which recordings were taken.
- PBSUM: total plant-body amount = sum of PB values (from PB00 to PB12) over all layers.
- SPBSUM: shrub plant-body amount = sum of shrub layer PB values (from PB01 to PB05).
- TPBSUM: tree plant-body amount = sum of tree layer PB values (from PB08 to PB12).
- PBHD: plant-body height diversity, calculated with the Shannon-Weaver formula, with Pi = the PB value of the i-th layer divided by PBSUM.

$$PBHD = -\sum_{i=1}^{s} Pi \bullet \ln Pi$$

Vegetation cover within the parks and land use outside the parks

The area of land covered with shrubs and taller trees within the urban park (= area of vegetation cover) was calculated using aerial photographs taken in 1998. Surrounding land uses were computed with GIS (geographic information systems) using topographic maps (1:2,500) and the aerial photographs. All land uses were classified into nine land-use types: rice field, dry field, orchard, woodland, bamboo forest, residential, road, wasteland and water. The proportion of each land-use type within 500 m of the edge of park was calculated for each urban park.

Data analysis

First the correlation of the size of the urban park and the area of vegetation cover to species richness was confirmed by regression analysis. In order to find patterns of bird species compositional change among the urban parks, TWINSPAN (two-way indicator species analysis; Hill 1979) was carried out. Bird species that were recorded in only one park were omitted from the analysis. As the result, 16 species (out of 21 species observed) were included in the analysis. The total number of recorded individuals by species was used in the calculation. The pseudospecies cut levels were defined as follows: 0, 1, 2–9, 10+ individuals. The groups that represented less than five parks were not split further.

Classification-tree analysis was performed to determine factors relating to compositional change. Classification-tree analysis has recently been used to determine major environmental factors of variation in plant/animal communities (e.g., De'ath and Fabricius 2000). Park grouping obtained by TWINSPAN was used as the dependent variable for classification-tree analysis. Explanatory variables of the analysis were the size of the park, the area of vegetation cover, PBSUM, SPBSUM, TPBSUM, PBHD and the proportion of different land-use types within 500 m.

Results

Regression analysis

As a result of bird censuses, 21 species and 1,825 individuals were recorded. The most species (13 species) were observed in the largest urban park (3.4 ha). No significant correlation between the number of species and the area of urban park was detected (p=0.1; Fig. 2), though a significant correlation between the number of species and the area of vegetation cover could be found (r=0.40, p<0.01; Fig. 3).

TWINSPAN

The result of the TWINSPAN classification is shown in Table 1 as a species-sample ordered table and in Fig. 4 as schematic diagram. The studied urban parks were classified into four types. The characteristics of each type were as follows: Type D (urban parks Nos. 12, 13, 14 and 15) comprised most large parks in the study area. Eastern turtle dove *Streptopelia orientalis*, pale thrush *Turdus pallidus*, Daurian redstart *Phoenicurus auroeus*, Japanese bush warbler *Cettia diphone* and black-faced bunting *Emberiza spodocephala* were characteristically observed in these parks (Table 1). It is generally known that these species, except *E. spodocephala*, prefer woodlands.

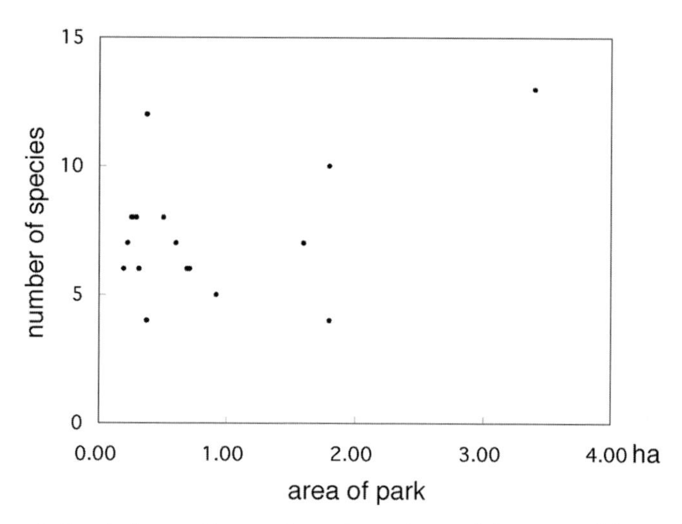

Fig. 2. Correlation between the number of species and the area of the park

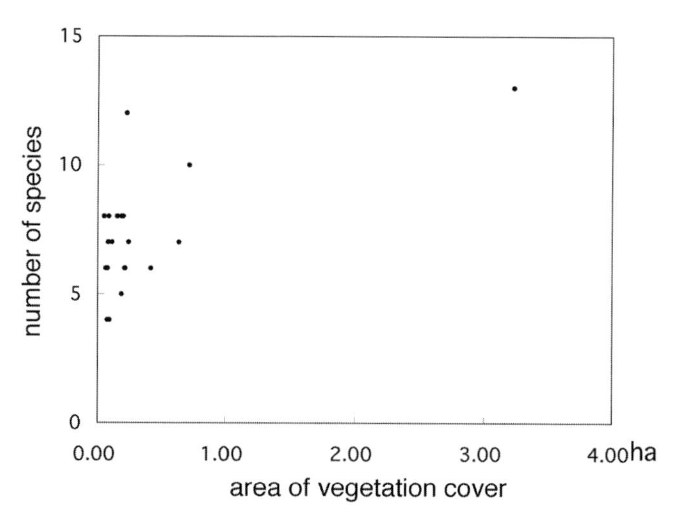

Fig. 3. Correlation between the number of species and the area of vegetation cover within urban parks

In type C parks (Nos. 1, 10, 16, 17 and 19) as well as in type D parks, the great tit *Parus major* was typically observed. Type C parks had smaller areas, but also high levels of tree vegetation cover. Grey starling *Sturnus cineraceus* was predominantly observed in type A (Nos. 6, 7, 8, 9 and 11) and C parks, and dusky thrush *Turdus naumanni* was mainly observed in type A. Type A (except one park) comprised the small parks located near large woodlands in the northwestern part of Nishinomiya City. Small and/or isolated urban parks were classified as type B (Nos. 2, 3, 4, 5, 18 and 20). In these parks, we recorded only bird species that tend to inhabit urban areas such as common pigeon *Columba livia* var. *domestica*, brown-eyed bulbul *Hypsipetes amaurotis* and mountain sparrow *Passer montanus*.

Table 1. Classification of species and parks by TWINSPAN. The output of TWINSPAN is a two-way table. It shows that parks are ordinated along the horizontal axis and species are ordinated along the vertical axis. All parks (and all species) are classified into a hierarchy of smaller and smaller (and more similar) groups by successive divisions. The divisions at each level of the hierarchy are indicated by lines of zeroes and ones below and to the side of the table. Where zeroes change to ones, a division is indicated (marked with lines). The numbers in the table show the pseudospecies cut level of each species.

Type	A					B						C					D				
	11	6	7	8	9	4	3	18	2	5	20	16	17	19	1	10	12	13	14	15	
Motacilla alba	-	-	-	-	-	-	1	-	1	-	1	-	1	1	-	-	-	-	-	-	00
Passer montanus	2	3	3	3	3	3	3	3	3	3	3	3	3	3	3	3	2	-	-	-	00
Sturnus cineraceus	3	2	3	2	2	-	-	-	-	-	-	1	2	2	2	3	-	-	1	-	00
Turdus naumanni	2	-	1	2	2	-	-	-	1	1	-	1	-	-	-	-	2	-	-	-	010
Hypsipetes amaurotis	2	2	2	3	2	2	2	2	2	2	2	2	2	2	3	2	2	3	3	3	011
Lanius bucephalus	1	-	-	-	-	-	-	-	-	-	-	1	-	-	-	-	-	-	1	-	011
Corvus macrorhynchos	-	-	1	1	2	-	-	-	2	2	2	-	1	-	2	2	-	2	-	2	011
Columba livia var. domestica	-	3	3	3	3	1	3	3	3	3	3	3	3	3	3	3	-	2	3	3	011
Streptopelia orientalis	-	2	-	2	1	-	1	-	-	-	-	-	-	-	-	-	2	-	1	1	10
Zesterops japonicum	2	2	2	1	2	-	-	1	-	2	1	2	2	2	-	2	3	2	2	2	10
Phoenicurus auroeus	-	-	-	-	-	1	-	-	-	-	-	-	-	-	1	1	1	-	1	1	110
Parus major	-	-	-	-	-	-	-	-	-	-	1	2	2	-	2	2	2	-	2	2	110
Turdus pallidus	-	-	-	-	-	-	-	-	-	-	-	-	-	-	-	-	2	1	-	2	111
Cettia diphone	-	-	-	-	-	-	-	-	-	-	-	-	-	-	-	-	-	1	1	1	111
Emberiza spodocephala	-	-	-	-	-	-	-	-	-	-	-	-	-	-	-	-	2	2	-	2	111
Eophona personata	-	-	-	-	-	-	-	-	-	-	-	-	-	-	-	-	-	-	2	1	111
	0	0	0	0	0	0	0	0	0	0	0	0	0	0	0	0	1	1	1	1	
	0	0	0	0	0	1	1	1	1	1	1	1	1	1	1	1					
	0	1	1	1	1	0	0	0	0	0	0	1	1	1	1	1					

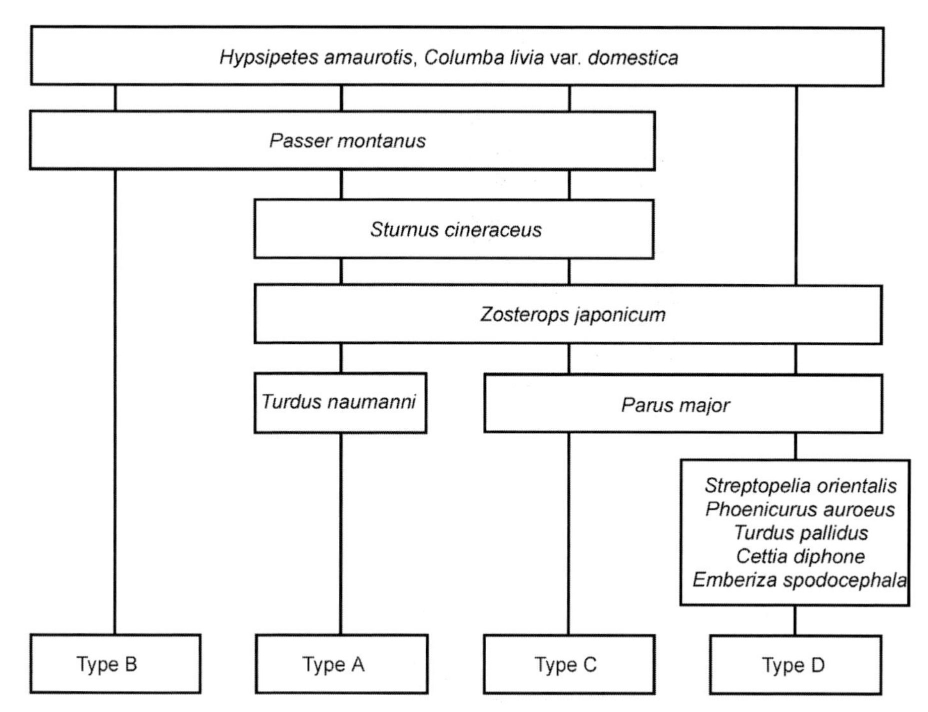

Fig. 4. Schematic diagram of typically observed species, classified by TWINSPAN into four park types

Classification-tree analysis

The results of the classification-tree analysis are shown in Fig. 5. First, urban parks were divided into two branches according to the proportion of surrounding woodland within 500 m. All urban parks of type B and two parks of types A and C were classified into a branch for which the proportion of surrounding woodland is 5.9% or less, while all parks of type D and the other parks of type A and C were classified into another branch with greater than 5.9% surrounding woodland. The former branch was divided by SPBSUM. The SPBSUM values of all urban parks of type B were 37.5 or less. The latter branch was divided by the area of vegetation cover. All urban parks of type D and one park of type C had an area covered by vegetation of more than 0.21 ha. The other branch (area of vegetation cover <= 0.21 ha) was divided by TPBSUM (splitting point is 1).

Using-layer selection

The using-layer selection is shown in Fig. 6. It was selected seven major species recorded more than 15 individuals. The heights at which birds were observed were classified into four layers: ground (0 m), herb and shrub (0–3 m), sub-tree (3–9 m) and tree (9+ m) layers. Most individuals of *Columba livia* var. *domestica* and *Passer montanus* were found on the ground. Most individuals of *Hypsipetes amaurotis*, *Zosterops japonicum*, *Parus major* and *Corvus macrorhynchos* used the sub-tree and tree layers.

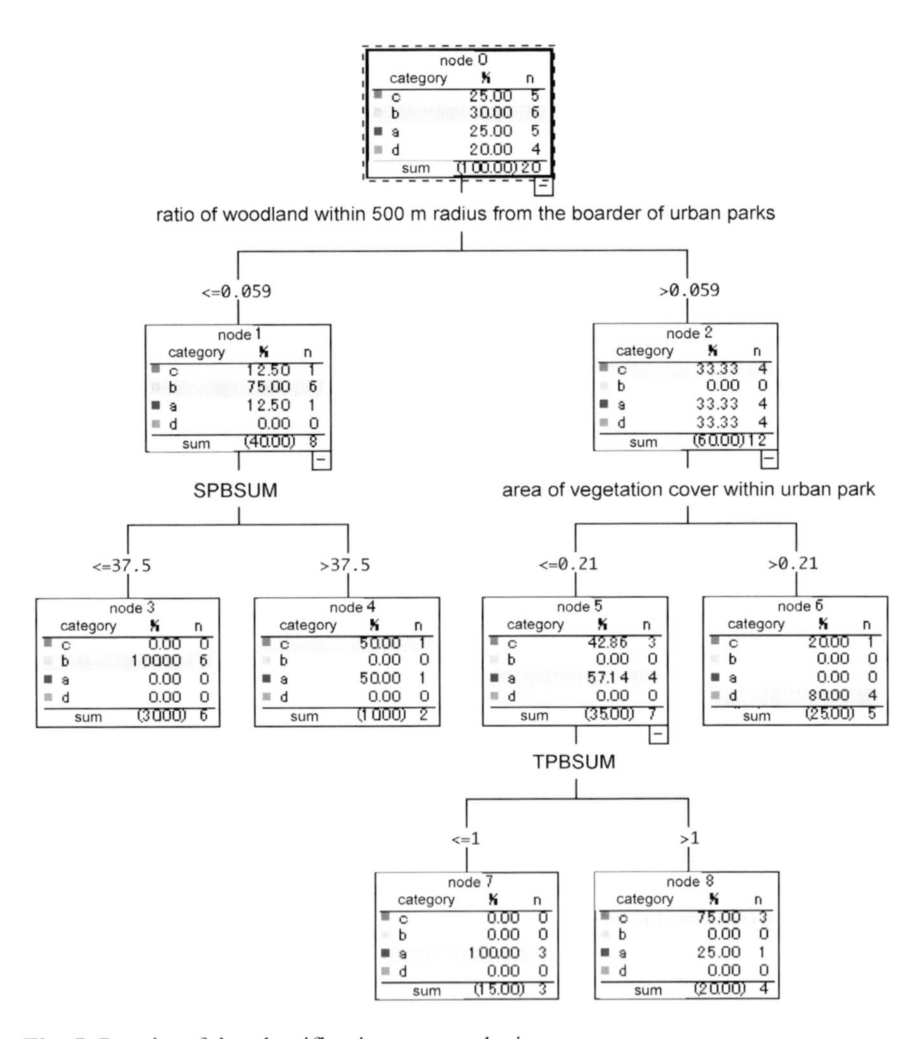

Fig. 5. Results of the classification-tree analysis

Discussion

Relationship between the number of species and the area of the park

Many studies (e.g., Martin 1983) have indicated a relationship between the number of species and the size of habitat for isolated habitats. However, I did not find a significant correlation between the number of bird species and the area of urban park, but there was a significant correlation between the number of species and the area of vegetation cover. The results of the classification-tree analysis also showed that the area of vegetation cover within an urban park is one of the important environmental factors. This means that the vegetation cover in parks is more important for many bird species than the size of the park. There are many small urban parks in Japanese cities, however most of them have no trees and shrubs.

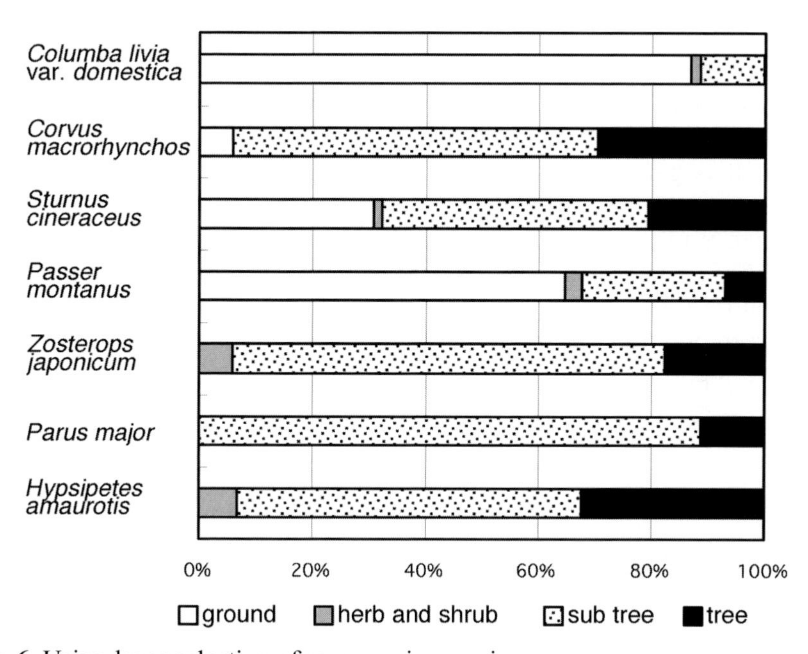

Fig. 6. Using-layer selection of seven major species

Vegetation structure and surrounding land use

The result of TWINSPAN and classification-tree analysis showed that the composition of bird species in urban parks was influenced by the proportion of woodland within 500 m of the edge of urban parks, the area of vegetation cover within urban parks, the presence of trees (at a height of more than 9 m) and the presence of herbaceous plants and shrubs (at heights of 0.12–3 m).

Many studies have suggested that the land-use pattern around habitats influences bird communities within the habitats, however only a few studies (e.g., Loman and Schantz 1991) have shown the result by field survey. In this study, classification-tree analysis showed that the most important splitting criterion was the proportion of woodland within 500 m from the edge of urban parks. If the proportion falls to less than about 6%, only a few bird species adapted to urban environments can inhabit an urban park. This value is very meaningful, because the proportion of green coverage (except farmland) of most cities in Japan is under 10%.

The relationship between the species richness of birds and vegetation structure has often been studied (e.g., Erdelen 1984; Yui and Suzuki 1987; Ichinose and Katoh 1998). While most studies indicated that forest-interior bird species prefer closed vegetation cover, Peason (1993) showed that the density of the shrub layer was also of similar importance for some wintering birds. In this study, the classification-tree analysis showed that the dense herb and shrub layer influenced the composition of wintering bird species. It is generally known that *Cettia diphone* and *Emberiza spodocephala* prefer dense shrub habitat. Also the presence of trees (at heights of more than 9 m) was selected as a splitting criterion in the classification-tree analysis. The urban parks of type A and C were divided by this criterion, and no *Parus major* was recorded in type A urban parks. Because *P. major* used only the sub-tree and tree layers, the presence of trees is one of the most important factors for *P. major* habitats.

Toward establishing ecological network planning for wintering bird species

The results of this study indicate the following points for ecological network planning in Japanese cities.

Protection of large core woodlands. Of course, large habitats are important for not only wintering birds but also breeding birds. In most Japanese cities, there are few large woodlands and the proportion of green coverage is lower than that of European cities. However, we have some large woodlands around cities such as this study area and some remnants of large woodlands and shrine and temple forests within urban area; we can establish ecological networks using such woodlands and forests as core habitats.

Improvement of the ratio of woodland around core woodlands. This study showed that the ratio of woodland within 500 m of urban parks was the most important environmental factor for bird species. In order to introduce forest bird species into the urban area, we have to increase the proportion of woodland near core habitats.

Conservation of tall trees. In cities, including this study area in central Japan, there are not many tall trees, because of the lack of rainfall during the summer season and typhoons. Also most trees in urban parks and along streets in Japan are strongly pruned in winter like "bonsai." The lack of tall trees affects the distribution of bird species that prefer woodland habitat. As the results of this study showed, all individuals of *Parus major* were recorded only in the sub-tree and tree layers. We must grow taller trees and select tree species suitable for the climate of this area.

Development of vegetation within urban parks. Many Japanese urban parks have no vegetation and are composed of bare ground and playgrounds. Even if there are trees, herbaceous plants and shrubs are usually pruned. Dense herb- and shrub-layer vegetation provides good habitats for some bird species. Such dense vegetation has been avoided in planning of small parks in urban areas because of security concerns. However, it might be possible to arrange vegetation that is suitable for both people and birds.

References

Bibby CJ, Burgess ND, Hill DA (1992) Bird Census Techniques. Academic Press, London

De'ath G, Fabricius KE (2000) Classification and regression trees: A powerful yet simple technique for ecological data analysis, Ecology 81:3178–3192

Erdelen M (1984) Bird communities and vegetation structure: I. Correlations and comparisons of simple and diversity indices. Oecologia 61:277–284

Haase R, Littel M, Lorenz W, Söhmisch R, Zehlius W (1992) Neuanlage von Trockenlebensräume. Bayerisches Staatsministerium für Ernährung, Landwirtschaft und Forsten, München

Hill MO (1979) TWINSPAN—A FORTRAN program for arranging multivariate data in an ordered two-way table by classification of the individuals and attributes. Cornell University, Ithaca

Ichinose T, Katoh K (1998) Factors influencing bird distribution among isolated woodlots on a heterogeneous landscape in Saitama Pref., Japan. Ekologia (Bratislava) 17:298–310

Kirby KJ (1994) Are habitat corridors conduits for animals and plants in a fragmented landscape? English Nature, Peterborough

Loman J, von Schantz T (1991) Birds in a farmland—More species in small than in large habitat island. Conservation Biology 5:176–188

Martin J-L (1983) Impoverishment of island bird communities in a Finnish archipelago. Ornis Scandinavica 14 :66–77

Nowicki P, Benett G, Middleton D, Rientjes S, Wolters R (eds) (1996) Perspectives on ecological networks. European Centre for Nature Conservation, Arnhem.

Pearson SM (1993) The spatial extent and relative influence of landscape-level factors on wintering bird populations. Landscape Ecology 8:3–18

Yui M, Suzuki Y (1987) The analyses of structure of the woodland bird communities in Japan. Journal of Yamashina Institute Ornithology 19:13–27

Nature Conservation, Forestry, Landscape Architecture and Historic Preservation: Perspectives for a Conceptual Alliance

Stefan Körner

Institute of Ecology, Technical University Berlin

Introduction

Administrative nature conservation in Germany, which is predominantly scientifically and ecologically based, must admit to a lack of public acceptance (Politische Ökologie 1995; Wiersbinski et al. 1998; Der Rat von Sachverständigen für Umweltfragen 2002; Körner et al. 2003). (Volunteer nature conservation organizations are not the subject here, nor are the many local nature conservation projects.) This lack of public acceptance is sparked by the widely held perception that the work of nature conservation is often limited to species conservation, and that, at least in its most common form, it leads to the interpretation of human uses of nature as "disturbances." Fischer (2004) has gone so far as to suggest that nature conservation has a fundamentally restrictive character: "It has come to see itself, at least here in Germany, predominantly as a restraint, a containment, as a repression of 'unchecked' human behavior, that only too obviously follows maxims [that describe humans as] separate from nature and enemies of nature" (ibid, p 25f). Particularly in the establishment of national parks, every permitted 'encroachment' into nature "is understood as a more or less grave, ultimately lamentable reduction from the norm-determined optimum, from the 'independent workings' of a nature insulated from human hands" (Fischer 2004a, p 224). Here normatively, a kind of nature is conceptualized "that is only complete nature, where human activities, even human presence is kept at a distance" (ibid, p 230). Although a restrictive nature conservation, for example, for the protection of white-tailed eagles, cranes and migrating falcons can certainly be justified, one should not wonder that this kind of nature conservation is understood by the public as excluding them from the landscape that they use everyday, and consequently from their *Heimat* (homeland) (cf. Stoll 1999; Bogner 2004).

Kowarik I, Körner S (eds) Wild Urban Woodlands.

This narrowly formulated understanding of nature conservation has historic roots in the so-called nature conservation in the narrow sense. It is expressed – as we will see in the example of Berlin's Südgelände – today in the development of urban-industrial abandoned areas and leads to the regulation of legitimate uses, even when the opposite is intended.

The examples in this chapter will show that new demands will be placed on nature conservation especially in the design of urban-industrial woodlands in urban contexts. These demands include unifying a non-restrictive nature conservation with the management of urban forests, recreation uses, and historic preservation in a landscape architectural, i.e. in a functional, satisfyingly designed, form. Satisfyingly designed means that the character, i.e. the uniqueness, of the post-industrial landscape in which these woodlands are found is not destroyed, but rather is further developed in a way appropriate to our time.

To achieve this demand, nature conservation can connect to a tradition that stands in contrast to its own narrow interpretation. There exists within the German tradition of *Heimatschutz* (the preservation of the homeland) a broader understanding of nature conservation that is more open to human uses of the cultural landscape (i.e. nature conservation in the broader sense). *Heimatschutz* invokes the even older tradition of *Landesverschönung* (the aesthetification of the land), in which the beauty of the landscape was seen as an expression of its usefulness (cf. Körner and Eisel 2003, p 12f). *Landesverschönung* represents an important link between a more broadly conceived nature conservation and forestry.

This revival of a broader understanding of nature conservation is initially opposed by the fact that the relevant disciplines of nature conservation, forestry, landscape architecture and historic preservation became differentiated in the past. However, they feature similar traditions, similar as well to *Heimatschutz* and *Landesverschönung*, that could be used in the urban-industrial context. There exists, for instance, within forestry, in the concept of forest aesthetics, a historic orientation not only toward wood production, but also toward the alliance of forestry with the *Landesverschönung* tradition of – one would say today, the landscape architectural way of – designing of forests for recreation.

Nature conservation worldwide has had as its aim, at least since the Convention on Biodiversity, the preservation of biological diversity. The German tradition of *Heimatschutz*, however, provided for the landscape architectural design of the landscape with both aesthetic and use-oriented perspectives in mind. In this way, landscape architecture, often sharply criticized today by nature conservationists as antiquated and culturally sterile and with its interests firmly oriented to the design of urban spaces (Mattern 1950; Bappert and Wenzel 1987; Kienast 1981), was closely connected to *Heimatschutz* and the broader understanding of nature conservation.

Fig. 1. Landschaftspark Duisburg-Nord

In turn, historic preservation is oriented predominantly toward the preservation of historic structures, and views the emergence of wild urban woodlands with reserve, because, for instance, as old rail yards become overgrown, the historic built structure is threatened (cf. Buschmann 2003). The preservation of *Verwilderung* processes (the processes by which land becomes increasingly wild) is seen, from this point of view, to be an approval of such decline. This is understandable to the extent that, as will be shown, the preservation of process does indeed seek to overcome the connection of traditional nature conservation with historic preservation and to assist 'pure' dynamic nature break free. It should be remembered that the most broadly understood approach to process protection within the German-speaking world has not only been shaped by ecological-scientific interests, but also significantly by cultural-historic interests, above all the preservation of character. The experiences with urban nature conservation illustrated below, which are in some ways distinct from the characteristic understanding of nature conservation, in conjunction with the adoption of the nature garden concept in landscape architecture, should broaden the perspective for a process-oriented historic preservation within urban-industrial woodlands.

All disciplines relevant to the management of urban-industrial woodlands feature cultural/design-oriented traditions that can both bring forth new, interesting solutions and be used for a broader, not exclusively restrictive nature conservation. The goal of this chapter is to study the historic connections between the individual disciplines in order to achieve an interesting synergism through the coordination of their currently unused potentials. A cooperation of this kind is of great interest in view of the lack of public acceptance described above. In the following, the possibilities for a conceptual alliance between nature conservation, forestry, landscape architecture, and historic preservation in the design of urban woodlands will be described.

Nature conservation

Although cultural interests are described in the *Bundesnaturschutzgesetz* (Federal Nature Conservation Act) at the same hierarchical level as ecological interests, there exists in Germany an influential movement that understands nature conservation exclusively as an ecological-scientific task. In connection with this understanding, the fallacy that objectively compelling maxims for political action would result from ecological data is widely prevalent. The trivial matter that ecological knowledge must, from a natural-scientific definition, be neutral and therefore must first be assessed in view of societal values in order to become actionable is often forgotten. But even then, when the difference between ecology as a natural science and as a societal field of endeavor is considered, cultural interests disappear from the field of view, because they aren't considered to be objective, and nature conservation is regarded solely from a technical perspective.

So, according to Erz (1986), a well-known representative of the federal nature conservation administration, nature conservation research, in the framework of its legally defined objectives, has the following fundamental tasks to fulfill:

- "Improvement of laws and administrative regulations
- Proper guarantee of their execution
- Determination of scientifically objective criteria for deciding individual cases
- Provision of rationales for citizen participation
- Determination of measures of success for all political, administrative and technical measures" (ibid, p 11)"

Ecology is suited to serve as a basis for scientifically objective conclusions (ibid, p 11), because one expects from it technically usable expert knowledge: "Ecological science has its goal (as is true for every science) to achieve, through research, an objective, true vision of these studied objects (functional elements in ecological systems, interdependency, etc.) and through this to produce a systematic knowledge in the form of generally applicable basic principles and general laws that remain the same for every examination using this method, independent of different observation methods of different people (subjectivity)" (ibid, p 12). In this sense, objective knowledge assigns observable results to theoretically determined classes (subsumption principle) and explains them causally through cause and effect relationships. The theories and observation propositions to be applied must be independently formulated in order to be generally and empirically, i.e. experimentally and intersubjectively, testable (cf. Popper 1972). This results in not only general relevance, but also transparency for political decisions.

In contrast, technology involves the most possible effective application of laws through engineering on the basis of defined social objectives (cf. ibid, p 52). Technical action is therefore, always use-oriented and rational. For this reason, the summary of nature conservation research formulated by Erz, which through scientific observations and certain resources in the political-administrative framework, aims to improve the implementation of nature conservation in the democratic process and wishes to implement specific desired states of nature and society, corresponds fundamentally to a technical, i.e. an instrumental, vision: Technical research, as an applied science, optimizes practical knowledge in order to achieve a greater efficiency in problem solving. A significant portion of the research in the field of nature conservation is concerned with the development and improvement of instruments of implementation.

In this understanding of nature conservation as a technology- and process-oriented, instrumental field of knowledge, which derives its general theories and laws from ecology, cultural questions are based on "subjective, social valuations" (Erz 1986, p 12), because they can not be generalized in the sense of the natural sciences. The aesthetic perception of character and beauty in the landscape, for example, is not reproducible independent of the observer and therefore is not universally valid in this sense. Thus an enduring dissatisfaction prevails that landscape beauty can not be evaluated objectively in the framework of landscape image analysis for the purpose of determining its value for recreation.

Based on this technical-instrumental understanding of the tasks of nature conservation and despite the legal equivalence of both ecological and cultural nature conservation in the *Bundesnaturschutzgesetz*, a strong tendency exists to only consider the scientifically actionable areas as objective and therefore, quietly to consider them the only socially necessary ones. This view was recorded from a planning theory point of view by Bechmann (1981) in his handbook about planning theory and methodology. This is, therefore, in no way an arbitrary, individual opinion of Erz, but rather in its paradigmatic expression it is a broad general view of nature conservation. (Through the development of landscape planning after World War II based on the National Socialist land management in Germany, additional references can further substantiate how cultural interests, seen as subjective, were intentionally removed from the understanding of the tasks of the instrumental, target-oriented view of nature conservation. This particularly German case will not be addressed here; cf. Körner 2001 p 77).

With the rejection of the cultural background of nature conservation, its historic connection to forest aesthetics, to landscape architecture and to historic preservation have mostly disappeared from the consciousness of the nature conservationist. This is all the more regrettable as the valuation of regional character described in the *Bundesnaturschutzgesetz* shows the connection to historic and therefore to cultural categories. The preservation and the contemporary design of the historical character of a landscape was the core of the broader understanding of nature conservation, upon which the synergy of interests mentioned in the Introduction could be established in the management of urban-industrial woodlands.

Nature conservation in the broad sense

In contrast to a more narrow understanding of nature conservation, Schoenichen (1942), for example, spoke of a nature conservation in the broad sense. With this, an active design of the landscape in the sense of modern landscape architecture was meant. This view of nature conservation is predominantly culturally motivated, in that the cultural landscape, or the *Heimat*, should be designed according to the requirements of humans uses, while protecting the character that is to be expressed in the image of the landscape. Landscape architecture in the sense of an old definition of *cultura* as the human utilization and development of nature through cultivation and construction of the landscape is therefore a constructive exercise in the structural sense; it brings the functional and aesthetic aspects of land use into harmony. Uses are not fundamentally 'disturbances' of a nature that is seen as intact, but rather activities that create culture or enrich the landscape. The beauty of the landscape should be a result of its usefulness.

This idea of connecting beauty with function goes back to an even older tradition: *Heimatschutz*, which Schoenichen classified as belonging to nature conservation in the broader sense, corresponds to the tradition of *Landesverschönerung* in the 19th century, which was not an aesthetic program as the name suggests, but rather an agriculture and forestry modernization program (cf. Däumel 1961). The field of forest aesthetics – as we will see – also belongs to this tradition. Functionality and the beauty that results from it were regarded as the expression of truth and goodness (ibid).

Process conservation as a nature conservation strategy

The strategy of current nature conservation that is oriented toward preserving a historic character has recently been criticized as 'unnatural' because it misjudges the evolutionary, dynamic character of nature and ultimately protects only the species and biotopes of the historic landscape (Scherzinger 1990, 1991, 1995, 1996, 1997; Reichholf 1994). Nature conservation, according to Reichholf, is fighting the wrong battle and is "a form of historic preservation": "It wants to preserve 'images of the landscape'. Anything that changes the familiar picture is reflexively fought against" (ibid). Here the historic preservation character of nature conservation is explicitly being questioned. This confirms the fears of the historic preservationists, who believe that acceptance of the processes of *Verwilderung* in urban-industrial spaces runs contrary to the interests of historic preservation. Process-oriented nature conservation wishes to protect nature as much as possible in a state where it can freely develop and does not address preserving the remnants of history.

Nevertheless, there are points of alignment between process conservation and historic preservation: First, it is not the case, at least in the most popular German understanding of process conservation, as described by Scherzinger who writes of the development of woodlands in the 'open landscape,' that a completely free development of nature is the goal. The goal, as in conventional nature conservation, is a 'proper' development of nature. This is oriented, furthermore, toward the historic character of the landscape, only in Scherzinger's understanding it is not the cultural landscape that is predominantly meant, but rather, within limits, the wild forest. In this way of thinking, current processes, such as the colonization of non-native species, are to be stopped because non-native species are not seen as an improvement but rather as a threat to native, i.e. traditional species diversity. Locally extinct species, in contrast, may be reintroduced at great expense (Scherzinger 1995).

Although an image of nature is constructed through Scherzinger's valuation of regional character that is essentially based on cultural criteria, he presents his view, as do many others, as purely ecologically based. It can be stressed however, that this form of process conservation, in contrast to the views of Reichholf, demonstrates a fundamentally historic dimension that permits a connection with historic preservation to be made. The historically determined species composition of an area would then have to be interpreted in light of the historic preservation character of a space (cf. Kowarik et al. 1998). This can be achieved with the appropriate work of successful contemporary landscape architecture in urban nature conservation.

Urban nature conservation

Urban nature conservation, as it was based significantly on the foundations of the Berlin urban ecology movement (cf. Wächter 2003, pp 91f), fundamentally follows the same goals as general nature conservation set by the *Bundesnaturschutzgesetz*. In this way, urban nature conservation can not initially be differentiated from nature conservation in the 'open' landscape. However, a more open view of present changes in nature development is taken in comparison to the process conservation orientation described by Scherzinger. The colonization of non-native species as an expression of contemporary change, for instance, is accepted because urban nature is substantially determined by the dynamic urban diversity of uses and because the colonization of non-native species is typical of this. Reichholf holds a similar position (1996, 1997). In contrast to Reichholf, however, in the urban nature conservation of the Berlin mold, the colonization of non-native species is not analyzed from a purely ecological perspective, but is also valued from a cultural-historic one (cf. Kowarik 2003).

Fig. 2. Former Zollverein coal mine

Through urban ecological research, it has become clear that the city presents a diverse and characteristic nature. Kowarik has divided this nature into four types. These four natures of the city represent broadly differentiated social uses and therefore, cultural forms with, in some ways, different historic dimensions. Nature of the first kind represents the remaining primeval wilderness, nature of the second kind agricultural nature, nature of the third kind horticulturally designed nature for which different eras and styles can be distinguished, and nature of the fourth kind which represents urban-industrial nature (cf. Kowarik 2005).

Based on the clear dependency of urban nature on urban uses, these uses can not be understood per se as 'disturbances' of an ideal condition of nature, particularly because this ideal condition, as it is usually understood in conventional nature conservation, pertains to be a rural pre-industrial condition that exists in cities only in remnants. In addition to the cultural-historical dimension, the social, use-oriented and livability dimensions of urban nature conservation are particularly emphasized: "The goal is *not* to set aside as large areas of the city as possible and to declare them as 'nature reserves' and where possible fence them in order to deprive the public of the use of these areas for recreation" (Sukopp and Kowarik 1988, p. 48). The goal of ecological urban design is, far more, to improve, to augment or to create anew quality in the variety of green spaces. "Species and biotope conservation is not an end in itself, but rather an explicit part of this human-oriented strategy" (ibid). It is usually said that, fundamentally, nature conservation – including the more narrow understanding – is there for humans. Ultimately what matters is exactly what this is understood to mean and whether that finds acceptance with the public. For Berlin urban nature conservation, for example, this social orientation has the result that the appearance of wear and tear in urban open-space vegetation due to intensive use is not seen as vandalism, but is tolerated (ibid, p 52). A moderate degree of disturbance is more often seen as increasing diversity (Kowarik 1993, p 19).

This view, which is in contrast to the characteristic understanding of nature conservation described in the Introduction in that it does not fundamentally interpret the human use of nature as the 'disturbance' of an 'intact' nature, has significantly influenced the concept for the Schöneberger Südgelände Nature Park (cf. Kowarik and Langer 2005). The concept for the Nature Park allowed for a distinction between the central nature conservation area where visitors must keep to the paths and the surrounding landscape conservation areas. In the latter, open access to an experience of nature as unregulated as possible and to free children's play was to be allowed in order to combine urban recreational uses with nature conservation in a heavily settled inner-urban area (cf. Kowarik and Langer 2005). Nevertheless in the course of the maintenance of the Nature Park increasing attempts have been made to keep visitors on the paths throughout the park and thereby to assert a more narrow, predominantly species conservation-oriented view of nature conservation (see Kowarik et al. 2004). The appropriateness of this maintenance policy is currently being discussed with the highest nature conservation authorities and the first signs of a change of thinking are apparent.

In addition to the readiness to see the use of urban open space not as the destruction of nature, the cultural-historic dimension of urban nature conservation results in a basic willingness to cooperate with historic preservation in order to preserve cultural heritage (cf. Kowarik et al. 1998). As well, cooperation with landscape architecture follows from its interest in urban quality of life and its acceptance of social functions of urban nature. The Berlin philosophy, in addition to the nature garden concept still to be discussed, has significantly influenced landscape architectural work with the spontaneous development of nature in urban-industrial spaces. First, however, the links with the tradition of forest aesthetics will be presented.

Fig. 3. Schöneberger Südgelände

Forest aesthetics

Because woodlands grow on many urban-industrial sites as a result of natural succession, it stands to reason that these areas can be seen as elements of forestry. In urban regions, however, these woodlands are not so much subject to forestry uses such as wood production, but rather take on recreation as their primary function. This makes necessary a new orientation of urban forestry that takes into account social uses of forests. This may be provided by its historic roots in forest aesthetics, which had a closer connection to the cultural-historic and more broadly based understanding of nature conservation and therefore also to urban nature conservation and to landscape architecture. These roots can be rediscovered through the design of urban-industrial forests.

Forest aesthetics as a component of forestry can basically be classified as a use-oriented design approach that has close connections to *Landesverschönung* and *Heimatschutz* and, therefore, to historic preservation as well (von Salisch 1911, p. 1). Von Salisch defined the art of forestry as an "art of necessity", in which the outer form of objects is derived from their function. In this way, forest aesthetics is not seen as a solely aesthetic problem and does not, therefore, belong to the fine arts. This does not, however, make it secondary in von Salisch's opinion because, since humans are rational beings, they can only ultimately be satisfied with functionality (ibid, p. 23).

In his book *Forstästhetik* (Forest Aesthetics), von Salisch begins the chapter "Applied forest aesthetics" with the classification of the most practical uses of soil (ibid, p. 193). He addresses the connection with historic preservation, i.e., with the historic preservation of nature, in the chapter "Foundations of forest aesthetics" and points out the close connection with *Landesverschönung* and with *Heimatschutz* (ibid, p. 12).

In addition to his description of effective uses of soil, von Salisch describes the components of beauty in forests. Here the close connection with the historic preservation of nature plays a particular role, as not only are the different tree species with their specific aesthetic characteristics listed, i.e., their growth, leaf and bark forms (ibid, p. 80f) but also special rocks and rock formations and historically meaningful trees are described both as natural monuments and as signs of earlier forms of forestry (ibid, p. 166).

It is, however, decisive that from the historic preservation component of forest aesthetics, it is not a conservational approach that is derived, but rather that, for von Salisch, the interests of land use take priority. Natural- and cultural-historical interests as well as aesthetic interests must therefore be integrated into wood production. Then only when wood production is not given priority, as is the case today in the woodlands of urban-industrial areas, can the cultural-historical and social dimensions of forestry be placed in the foreground. Social dimensions for von Salisch include, for example, recreation for children and youth. Von Salisch quotes Riehl, the first German folklorist, with the words: "A village without a forest [is] like a city without historic architecture, without monuments, without art collections, without theaters and concerts, in short without leisurely and aesthetic stimuli. The forest is the playground of the youth, often a festival hall of the elderly. Does this not weigh at least as heavily as the economic question of the wood?" (Riehl quoted in ibid, p. 196). Von Salisch described the recreation function of forests as well, which is the understood purpose today of forests in conurbations and, therefore, also of forests in urban-industrial areas.

Landscape architecture

As in forestry, landscape architecture also has a tradition which may bear fruit in the design of urban-industrial spaces. Landscape architecture demonstrates – as we have seen – a close connection with urban nature conservation. At first, landscape architecture was marginalized in the universities and also in the public perception because of the development of the technical-instrumental vision of nature conservation described above. Because only scientific questions were seen as objective and able to be legitimated socially, a conflict arose early in Germany between ecologically oriented nature conservation and culturally motivated landscape architecture. Landscape architecture sees the individual design of spaces as a cultural responsibility and feels committed to a more artistic understanding of its responsibilities. The understanding of the tasks of administrative nature conservation that emerged after World War II were criticized as culturally unproductive, and the development of the landscape into a modern landscape for living with a contemporary and artistically satisfying design was demanded (cf. Mattern 1950). Artistic in this concept does not mean that landscape architecture was to be understood as one of the fine arts, but rather, like forest aesthetics, as an applied art. As an architectural discipline, landscape architecture designs the landscape according to the interests of human uses. Because of its artistic component, landscape architecture was long considered subjective and therefore irrational (cf. Milchert 1996) in the German universities, leading to its marginalization.

The landscape architecture tradition that is relevant to the design of urban-industrial spaces was already set out in the *Heimatschutz*. Despite its original orientation, expressed by Rudorff (1897) in terms critical of civilization, the idea of *Landesverschönung* was further developed within the *Heimatschutz* into the union of beauty and function. It no longer described only agricultural and forestry melioration programs, but also incorporated industrial modernization. Then it was not only the case that the incorporation of industrial works within the characteristic landscape – perhaps through the plantings around them, through the use of natural stone facades or as in the careful siting of the autobahn to the existing topography (cf. e.g. Seifert 1941) – was seen as a cultural accomplishment but it was understood that the industrial structures could themselves shape the landscape. Lindner (1926), who significantly influenced the relationship of *Heimatschutz* to industry and technology with his book *Ingenierwerk und Naturschutz* (Engineering and Nature Conservation), pointed out that the building projects of what was then modern industrial architecture would generally, in fact, be seen as ugly, but they were characteristic, i.e. they had a uniqueness. He not only attributed to high-tension power lines the ability to give a monotonous landscape a "new allure" (ibid, p 88), but also argued that the cinder heaps of the heavy industry in the Ruhr were a valuable contribution to the regional character: "The powerful rubble – and one would not want to do without the cinder heaps of the Ruhr, and we recognize them as an inseparable accompaniment to the mining and steel-mill industries and include them automatically in our vision of the homeland as a component of sentimental value" (ibid, p 92).

The landscape architecture of today in urban-industrial spaces represents a contemporary interpretation of the tradition of *Heimatschutz* and forest aesthetics. Consequently, the industrial structures as well as the forests that have arisen in the course of succession are included as elements of cultural-historical significance in designs. Within landscape architecture, the work with spontaneous natural development and therefore, with the development of woodlands, is influenced not only by urban nature conservation but also by the nature garden concept and by the tradition of urban open-space planning that allows for the use of abandoned sites. This tradition relates back to the so-called *Kasseler Schule* (Kassel School) (cf. e.g. Böse 1981).

The nature garden concept in landscape architecture

The nature garden concept, as it is described by the Swiss author Schwarz, consists of promoting nature conservation in the garden by developing the natural biotopes that one would find in the 'open' landscape as much as possible in the city, such as hedges, wildflower meadows, ponds, etc. The urban open spaces should be made into compensatory spaces for threatened native species that find no place within the landscape of intensive agriculture (see Schwarz 1980). On the other hand, the nature garden concept was greatly influenced by the Dutchman LeRoy, who did not see abandoned areas and gardens that have gone wild as 'nature conservation areas,' but primarily as places for the free development of nature and for human activities, especially of children. He worked architectonically, building dry walls, paths and minimal pioneer habitats out of building rubble (cf. LeRoy 1978; for a summary see Andritzky and Spitzer 1981).

This perspective influenced the design philosophy of the landscape architect Peter Latz, as can be seen, for example, in the dry walls and planting beds of building rubble designed by him for the Hafeninsel (the Harbor Island) in Saarbrücken (cf. Latz 1987). The Swiss landscape architect Dieter Kienast also grappled intensively with the nature garden concept. He seized the innovative ideas that were bound up in the idea, but did not submit to the nature conservation didactic of Schwarz's variation. Kienast insisted on the use-oriented garden, which he saw as more appropriate for a livable space outside one's home, as opposed to designs that imitated nature (cf. Hülbusch 1978 as an author of the *Kasseler Schule*). Kienast described urban abandoned areas which are not subject to any restriction as 'natural' archetypes of the 'livable' open spaces: "LeRoy and others have built or initiated such spaces together with those that are affected by them – the users…In this type [of nature garden] there is no imitation; the artifact is allowed to clearly emerge. Cultivation, construction and nature are not antipodes in their manifestations; they allow the autonomy of the other parts to stand or to become more striking in their manifestation" (Kienast 1981).

Kienast offers an ideal example of a use-oriented open space in Berlin's Gleisdreieck: "As one of the best examples of such a distinctive area, I got to know the Gleisdreieck in Berlin a few weeks ago. The cultural structures of the tracks and rail yard are still clearly recognizable within the spontaneous vegetation despite forty years of mostly undisturbed neglect. Soil formation, microclimate, water and light availability, and intensity of use have lead to the development of diverse and differentiated plant communities…. Newer built interventions come, above all, from children. Artistry and naturalness are concentrated here into an exhibition, whose preview has already been held in the everyday world. What can we learn from this? The work with history among other things and the ability to recognize quality where it exists" (ibid, p 42). For this reason, abandoned areas are often more important than official parks and are able to support landscape architectural design as a cultural expression quite well through the preservation of spontaneous vegetation and historic materials (ibid, p 42f).

Fig. 4. Gleisdreieck Berlin in 1981 (photo: Ingo Kowarik)

The achievement of entirely prosaic, functional requirements in open spaces, such as the design of entrances, the siting of paths and the placement of benches, is then a functional organization of open space that not only improves the possibilities for use, but also, through the carefulness of the interventions, avoids overwhelming the historic and natural structure. The concept for Berlin's Südgelände envisioned, therefore, spare interventions through the careful reuse of old structures (cf. Kowarik and Langer 2005); the rail lines, for example, were made into pathways and the main raised walkway through the nature conservation area was constructed over a set of tracks.

From the point of view of landscape architecture, the design of the urban-industrial nature of the Ruhr has to do with more than culturally aware design: this design has political components as well. Kienast describes the profession of landscape architecture (and nature conservation) as hostile to cities (ibid, p 106). This hostile position should be overcome as it is demonstrated through the design of former industrial zones that urban-industrial civilization can also produce meaningful cultural landscapes. Latz (1999, p 14) attributes to landscape architecture the task of developing landscape concepts that change the contemporary understanding of nature and compel a new discussion about nature in the city. In this way through design it should be made clear that cities also produce a diverse and characteristic nature; cities provides for a 'good' existence (cf. Latz 1999a); the urban way of life should be given value.

Fig. 5. Landschaftspark Duisburg-Nord

It is not the value of nature that epitomizes this life-style, but rather the value of urbanity: The city stands as a place of cosmopolitan and democratic culture (cf. Bappert and Wenzel 1987) that is mirrored in the specific urban-industrial nature. It is the authentic urban nature, which does not, as with Schwarz, imitate pre-industrial, rural conditions. The important status of urban-industrial nature for landscape architecture and urban nature conservation allows the enormous birch–buddleia stands of the Landschaftspark Duisburg-Nord to be incorporated into the design concept as the black locust stands were in the Südgelände in Berlin.

Additional landscape architectural design strategies

In addition, landscape architecture is involved in making urban-industrial spaces readable as landscapes, i.e. as large-scale works of nature and culture. Thus in the Landschaftspark Duisburg-Nord, typical elements of idyllic cultural landscapes and gardens were incorporated within the heavy-industry context. A kind of cottage garden with boxwood (*Buxus sempervirens*) and hydrangeas was laid out in a former ore bunker and water lilies were planted in the reservoirs (cf. Körner 2003, p 98f).

Fig. 6. 'Cottage garden' in a former ore bunker

As well, groves and allées were made use of as a well-known medium for ordering the landscape (cf. Rebele and Dettmar 1996, p 129). For this reason, Peter Latz placed columnar black-locust trees that could be seen from a great distance on a former waste dump in the Landschaftspark Duisburg-Nord (for further formal design concepts, see Grosse-Bächle 2005; Henne 2005).

All in all one can say that, through a new type of nature that includes the processes of *Verwilderung* and whose development is ultimately made possible through the decline of former heavy industry, a new form of landscape has arisen. This has been integrated, by landscape architecture, into recognized landscape design topoi. In this way, the development of spontaneous nature and the preservation of historic structure are not seen as contradictory. It is not a conservational idea of historic preservation and certainly not a reconstructive one that is pursued. Rather the preservation of history, such as historic traces of use, is bound up into a landscape architectural design approach that aims to organize the everyday usability of open space and express an authentic, aesthetic whole. Ultimately the *Landesverschönung* ideal of the union of beauty and function will once again be aspired to.

Fig. 7. Scarred bunkers in the Landschaftspark Duisburg-Nord

There is a thread relating to historic preservation that joins nature conservation in the broad sense, forest aesthetics, and landscape architecture – it is the idea of a historically unique character. All three fit into the traditions of *Heimatschutz* and *Landesverschönung*. It has been shown however, that process conservation distances nature conservation from its own historic preservation heritage, although the category of unique character is still somewhat recognized and therefore so is the cultural-historical dimension. Based on this understanding of character in urban nature conservation then, nature is differentiated into different types and thereby an open connection to landscape architecture and to historic preservation is fashioned. The idea of combining nature conservation with historic preservation is, therefore, not new.

In the following, it will furthermore be demonstrated that an approach exists as well in historic preservation theory that understands a freely developing nature as a central component of modern historic preservation values. This is also not new (cf. Mörsch 1998). How process conservation that is culturally aware and historic preservation that is process-oriented can complement each other will also be discussed.

Historic preservation in urban-industrial spaces

The presence of spontaneous nature in urban-industrial spaces allows a new type of landscape to emerge that is to be further designed. The decline of former heavy industries, however, stands in the way of the development of a new landscape. The history of these landscapes can be interpreted as a story of decline. Acceptance of this decline poses the threat that valuable industrial structures of historic preservation value may be lost.

Huse (1997) sees the central message of these historic industrial sites as the expression of the triumph and decline of industry, its earlier subjugation of nature and its current powerlessness. Industrial architecture, according to Huse, was normally constructed to be used for one to two generations; constant change through ongoing rebuilding was characteristic of its use. This dynamism continues under new circumstances, namely as decay, after the use has been abandoned. What would represent a need for quick action to a preservation-oriented historic preservationist, is, for Huse, an expression of the aura of a historic object and worthy of conservation. Correspondingly he positions himself as opposed to conservational approaches and concludes that historic preservation in urban-industrial areas is only conceivable as a process-oriented action (ibid, p 88).

Fig. 8. The growth of nature and the decline of industry

Every conservational strategy would quickly come up against its limits, so that only a concept of controlled decline would be conceivable. This, however, currently has no scientific foundation (ibid, p 89ff). A process-oriented historic preservation of this kind would ideally complement a process-oriented nature conservation. The starting point for this lies in the historic preservation theory of Riegl, one of the most significant theoreticians in the field, whose work is relevant here above all in view of the relationship between historic preservation and nature conservation (cf. Mörsch 1998, p 94).

The historic preservation values of Riegl

Riegl distinguishes, in addition to *Gegenwartswerten* (values of things in the present), such as the actual use value of an object, different historic preservation values as *Erinnerungswert* (the value of things recalled). He describes especially the preservation of art and history (Riegl 1903, p 144). "According to the generally used definition, a work of art is every touchable and seeable or hearable work of man that shows a particular artistic value, an object for historic preservation is every such work that possesses historic value". The historic value is described by Riegl comprehensively: "We call historic everything that once was and today no longer is; according to the most modern concepts, we can add to this the further opinion that that which once was can never be again and everything that existed forms an irreplaceable and unmovable link in the chain of development" (ibid, p 145). Historic value can, fundamentally, apply to everything: "According to modern concepts, every human activity and every human skill of which testimony or tidings have been preserved by us, can, without exception, lay a claim to historic value: every historic event counts for us, in principle, as irreplaceable" (ibid). Because one cannot preserve everything, one should give one's attention to "especially conspicuous stages in the path of development of a particular branch of human activity" (ibid).

A historic work of art is, then, "an essential link in the developmental path of art history. The 'historic work of art' in this sense is actually a monument to art history, its value from this perspective is not 'artistic value' but rather 'historic value'" (ibid, p 146). The result of this is that the distinction between historic works of art and monuments of history is unfounded, as the former is included in the latter. All historic works of art of earlier times or at least all artistic periods must then have the same value. Clearly, however, in addition to an art-historic value a purely artistic value exists that is independent of the place of the work of art in the historic chain of development (ibid).

On the other hand, when considering the artistic value, the status that an object has is relevant: If it is considered to stand the test of time from an aesthetic perspective (however uncertain this basis may be), then the object has an objective status. If, on the other hand, the artistic value is perceived by a contemporary observer, the object's status is subjective and consequently relative – and relates to a modern ambition for art (Kunstwollen) and to the present. The artistic value is then no longer an Erinnerungswert, but rather a Gegenwartswert, so that therefore there is also no longer any historic value. Then one should, in historic preservation, sensibly only speak of historic monuments (ibid, p 146ff).

The interest in the artifacts of the human past and their historic value has in no way been exhausted. This interest is not directed at the object "in the condition in which it originated," but rather at "imaginings of the time that has passed since it originated, made manifest in the traces of age." This value draws its meaning from the fact that it awakens in "modern man the perception of the compulsory cycle of growth and death" (all quotes ibid, p 150). This, to Riegl, is *Alterswert* (value achieved through the accumulation of age). When historic preservation approaches nature conservation through the concept of *Alterswert*, when it values the process of aging, of death, and of becoming one with nature, when ultimately it values historic objects not as the works of humans, but rather as works of nature (Mörsch 1998, p 94), then it is clear that it is not a matter of a conservational nature conservation, but rather a process-oriented one.

The main point, however, is that the distinction between preserving man-made artifacts and preserving nature does not make sense, because not only does the aura of an urban-industrial site emerge from the impression of a union between the two, but also because nature itself has a cultural character, in this case, an urban-industrial one (cf. also Huse 1997, p 93). Through a dynamic understanding of historic preservation, *Alterswert* can play a special role in determining the value of a historic industrial site because the traces of time present in a rotting steel beam or in scarred bunker walls become just as clear as those present in the development of a specific spontaneous nature.

Alterswert in the concept of Riegl

Similar to historic value, *Alterswert* is an *Erinnerungswert* and not a *Gegenwartswert*. For this reason, *Alterswert* develops when a historic object has no practical meaning, when it has no more use value. Historic churches or residences may always continue to have use value, because they fulfill a need in the present.

Respect for the effects of nature is, according to Riegl, the culmination of the modern historic preservation movement because this value corresponds to a fundamental condition of human existence:

"Es ist vielmehr der reine, gesetzliche Kreislauf des naturgesetzlichen Werdens und Vergehens, dessen ungetrübte Wahrnehmung den modernen Menschen vom Anfange des 20. Jahrhunderts erfreut. Jedes Menschenwerk wird hierbei aufgefaßt gleich einem natürlichem Organismus, in dessen Entwicklung niemand eingreifen darf; der Organismus soll sich frei ausleben und der Mensch darf ihn höchstens vor dem Absterben bewahren. So erblickt der moderne Mensch im Denkmal ein Stück seines eigenen Lebens und jeden Eingriff in dasselbe empfindet er ebenso störend wie einen Eingriff in seinen eigenen Organismus. Dem Walten der Natur, auch nach seiner zerstörenden und auflösenden Seite, die als unablässige Erneuerung des Lebens aufgefasst wird, erscheint das gleiche Recht eingeräumt wie dem schaffenden Walten des Menschen" (ibid, p 162).

This quote signifies that humans are moved by the cycle of nature. Each artifact is perceived as a freely developing organism in whose development intrusion is forbidden. The sole possible course of action consists of preventing the organism's extinction. The modern human sees a part of his own life in a historic object and every intrusion into that object is felt as an intrusion into his own organism. The workings of nature, including its decomposing and destructive sides, have the same rights as the constructive activities of humans. This includes, as a consequence, a consciousness of one's own morality as an awareness of one's own mortality. The dignity of the historic object is determined, therefore, by the effects of nature that can be observed in it. In this way, for Riegl, human altruism finds fulfillment in respect for nature, which he saw as a trait particularly of the Germans (ibid, p. 162). Despite this nationalistically based interpretation, the significance of *Alterswert* lies in the fact that it can be understood as a democratic value. *Alterswert*, in contrast to historic value or artistic value, can be perceived by everyone, not just by experts (ibid, p. 164).

With this in mind, the decay of historic objects is, for Huse, no loss of historic value, but rather a remarkable demonstration of the contemporary meaning of *Alterswert*, which – as the discussion of landscape architectural work with urban-industrial spaces shows – can be incorporated into appropriate design concepts. For this reason, Huse characterizes Latz's concept for the Landschaftpark Duisburg-Nord as a stroke of luck for historic preservation (cf. Huse 1997, p 94).

Consequences for landscape architecture

To avoid misunderstandings: this paper is not making the case for an uncontrolled decay of industrial architecture. It is, rather, a plea that the development of nature in former industrial regions be interpreted not only as a sign of decline, but also as an integral component of the historic preservation value of historic industrial structures. It will be shown, before the backdrop of the latest discussions in nature conservation about process-oriented approaches, that these can be combined with process-oriented approaches to historic preservation, which can do justice to the character of former industrial areas as landscape ensembles in a perhaps unfamiliar way.

This kind of understanding of historic preservation as a complement to process conservation, however, breaks a taboo. According to Huse, for too long a belief in the restoration and the conservation of original objects has been imparted to the historic preservation movement. The effects of time and the limitations of conservation practice, i.e. the changeability of the object of conservation, which is an everyday matter in historic garden preservation, have been repressed (cf. Huse 1997, p 89).

In nature conservation on the contrary, significant doubts emerged in the 1980s about conservational practice, which led to the development of process-oriented approaches. If these were partially explicitly directed against a historic preservation practice in nature conservation, they nonetheless cultivated a direction, in combination with urban nature conservation and a socially oriented variation of the nature garden concept, that sees no sense in the separation between nature and the work of humans because nature is culturally influenced to the core of its being. Despite the resulting basic willingness to integrate wild areas into the design of urban-industrial open space, these spaces are arrayed with typical landscape elements in such a way that they can be read as cultural landscapes. The truism follows for work in urban-industrial landscapes that urban wild woodlands are not sensible everywhere and that they will most likely be accepted if they are maintained in a grove-like state and when they alternate with usable, open areas of wildflower meadows or shrubs that are suitable for recreation (cf. also Rink 2005; Jorgensen et al. 2005). This illustrates how appropriate the principles for the design of woodlands described in von Salisch's *Forstästhetik* are for today. They must be newly interpreted in specific urban contexts. The pragmatic, sensible result will not be a true wilderness, but rather a mosaic of woodlands with open areas that are more or less carefully maintained.

Examples such as the Landschaftspark Duisburg-Nord and Berlin's Südgelände exist; their usability, public acceptance and natural development must be studied. For other spaces, such as Berlin's cemeteries, where changing demographics and evolving burial rituals are making open space available, the Jewish cemetery in Berlin-Weißensee can offer guidance. It demonstrates impressively that, through *Verwilderung,* a fascinating combination of culture and nature can arise, in which the gravestones as historic markers provide cultural signs that stand in contrast to the growing wilderness. In the cemetery, the motif, described within *Alterswert* by Riegl, of transience as a component of the modern historic preservation movement takes shape vividly side by side with high ecological value and high suitability for quiet recreation.

Fig. 9. Jewish cemetery, Berlin-Weißensee (photo: Ingo Kowarik)

An additional relevance of such spaces may arise from the discussion being held in Germany about the 'Bambi syndrome' and about the establishment of wild spaces for experiencing nature, in which unregimented children's play is to be possible (cf. Schemel 1997). The term 'Bambi syndrome,' coined by the pedagogic scientist Brämer, refers to the fact that the environmental education of, above all, youth presents nature as fragile, threatened and therefore, worthy of respect and conservation, while simultaneously infantilizing and trivializing it (Brämer 1998). Regardless of whether or not this image of nature is even ecologically, i.e., scientifically based, its conveyance leads, as Brämer determined, to a moralistic self-exclusion from nature, because everything that young people do for fun in nature, such as hiking, having parties, or camping is internalized as the destruction of nature. This has the paradoxical effect of increasing the alienation from nature through ecological pedagogy rather than eliminating it. Trommer (1992, p 135) discusses this catastrophe didactic of environmental education, which can lead to impotent anger, doubt, hopelessness and fear of the future.

In abandoned industrial areas particularly, the message that nature is delicate and in need of protection comes across as almost absurd because the fascinating nature that exists there has arisen despite substantial environmental destruction. The destruction is, in fact, a prerequisite. Environmental pedagogy in this case must take note of the conquest of a destroyed landscape by an often surprisingly flexible and suitable nature and must bring out the sensuous and – if you like – hope-giving appeal of this process.

A dynamic, process-oriented unrestricted nature development can correspond, in contrast, to a more relaxed and free recreation and may be a reason why, in recent landscape architecture, work is increasingly being done with the processes of vegetation development (cf. Grosse-Bächle 2005). The examples of Grosse-Bächle illustrate that the fascination with spontaneity and unrestricted growth as a general theme of contemporary landscape architecture lies deeper than the significance of *Alterswert* in the interpretation of industrial historic structures. This fascination may originate with the ancient meaning of nature as *physis*: "In ancient times and in the Middle Ages, nature was 'physis', i.e., the bearing of fruit, the blossoming, the spontaneity, the self-creating order, to which we belong." There are elements incorporated in this that we can bring together in everyday thought, but not when we talk about nature as scientists. Nature, in the modern understanding shaped by the natural sciences, is "that which obeys the laws (of nature)" (Kant). The "spontaneity" of the premodern *physis* implies freedom, an absence of rules or laws. At the same time, it implies as well that something is unfolding in accordance with laws and these are such laws as we can not avail ourselves of, let alone regulate (Trepl 1992, p 53). The premodern understanding of nature is aligned in abandoned industrial areas with the estimation of *Alterswert* as the high-point of the modern historic preservation movement. *Alterswert* stands, ultimately, for a cosmological ordering of the eternal cycle of growth and death.

Acknowledgements

For discriminating critique and wide-reaching suggestions, I would like to thank Ingo Kowarik. And for the translation I would like to thank Kelaine Vargas.

References

Andritzky M, Spitzer K (eds) (1981) Grün in der Stadt. Von oben, von selbst, für alle, von allem. Rowohlt Verlag, Hamburg

Bappert T, Wenzel J (1987) Von Welten und Umwelten. Garten und Landschaft 97(3):45–50

Bechmann A (1981) Grundlagen der Planungstheorie und Planungsmethodik. UTB, Bern Stuttgart

Bogner T (2004) Zur Bedeutung von Ernst Rudorff für den Diskurs über Eigenart im Naturschutz. In: Fischer L (ed) Projektionsfläche Natur. Zum Zusammenhang von Naturbildern und gesellschaftlichen Verhältnissen. Hamburg University Press, Hamburg, pp 105–134

Böse H (1981) Die Aneignung städtischer Freiräume. Beiträge zur Theorie und sozialen Praxis des Freiraums. Kassel, Selbstverlag.

Brämer R (1998) Das Bambisyndrom. Vorläufige Befunde zur jugendlichen Naturerfahrung. Natur und Landschaft 73(5):218–222

Buschmann W (2003) Zwischen Reliktschutz und ganzheitlichem Flächendenkmal. Die Entwicklung der Industriedenkmalpflege in den letzten 20 Jahren. International Conference Abstracts "Wild Forests in the City. Postindustrial Urban Landscapes of Tomorrow" Dortmund. 16.-18. Oktober 2003

Däumel G (1961) Über die Landesverschönerung. Hch Debus, Geisenheim/Rheingau

Der Rat von Sachverständigen für Umweltfragen (2002) Für eine Stärkung und Neuorientierung des Naturschutzes. Sondergutachten. Ulmer, Stuttgart

Erz W (1986) Ökologie oder Naturschutz? Überlegung zur terminologischen Trennung und Zusammenführung. Berichte der Akademie für Naturschutz und Landschaftspflege 10:11–17

Fischer L (2004) Projektionsfläche Natur. Zum Zusammenhang von Naturbildern und gesellschaftlichen Verhältnissen. Einleitung. Hamburg University Press, Hamburg, pp 11–28

Fischer L (2004a) Natur – das Seiende jenseits von Arbeit. Reflexionen über eine neuzeitliche Grenzziehung. In: Fischer L (ed) Projektionsfläche Natur. Zum Zusammenhang von Naturbildern und gesellschaftlichen Verhältnissen. Hamburg University Press, Hamburg, pp 223–259

Grosse-Bächle L (2005) Strategies between Intervening and Leaving Room. In: Kowarik I, Körner S (eds) Urban Wild Woodlands. Springer, Berlin Heidelberg, pp 231–246

Henne SK (2005) "New Wilderness" as an Element of the Peri-Urban Landscape. In: Kowarik I, Körner S (eds) Urban Wild Woodlands. Springer, Berlin Heidelberg, pp 247–262

Hülbusch IM (1978) Innenhaus und Außenhaus. Umbauter und sozialer Raum. Kassel, Selbstverlag

Huse N (1997) Unbequeme Baudenkmale. Entsorgen? Schützen? Pflegen? CH Beck, München

Jorgensen A, Hitchmough J, Dunnett N (2005) Living in the Urban Wildwoods: A Case Study of Birchwood, Warrington New Town, UK. In: Kowarik I, Körner S (eds) Urban Wild Woodlands. Springer, Berlin Heidelberg, pp 95–116

Kienast D (1981) Vom Gestaltungsdiktat zum Naturdiktat – oder: Gärten gegen Menschen? Landschaft und Stadt 13(3):120-128

Körner S (2001) Theorie und Methodologie der Landschaftsplanung, Landschaftsarchitektur und Sozialwissenschaftlichen Freiraumplanung vom Nationalsozialismus bis zur Gegenwart. Landschaftsentwicklung und Umweltforschung 111. TU, Berlin

Körner S (2003) Postindustrielle Natur – Die Rekultivierung von Industriebrachen als Gestaltungsproblem. In: Genske D, Hauser S (eds) Die Brache als Chance. Ein transdisziplinärer Dialog über verbrauchte Flächen. Springer, Berlin Heidelberg, pp 73–101

Körner S, Nagel A, Eisel U (2003) Naturschutzbegründungen. Landwirtschaftsverlag, Bonn – Bad Godesberg

Körner S, Eisel U (2003) Naturschutz als kulturelle Aufgabe – theoretische Rekonstruktion und Anregungen für eine inhaltliche Erweiterung. In: Körner S, Nagel A, Eisel U (eds) Naturschutzbegründungen. Landwirtschaftsverlag, Bonn – Bad Godesberg, pp 6–49

Kowarik I (1993) Stadtbrachen als Niemandsländer, Naturschutzgebiete oder Gartenkunstwerke der Zukunft? Geobot. Kolloq. 9:3–24

Kowarik, I. 2003: Biologische Invasionen: Neophyten und Neozoen in Mitteleuropa. Ulmer Verlag. Stuttgart

Kowarik I (2005) Wild urban woodlands: Towards a conceptual framework. In: Kowarik I, Körner S (eds) Urban Wild Woodlands. Springer, Berlin Heidelberg, pp 1–32

Kowarik I, Langer A (2005) Natur-Park Südgelände: Linking Conservation and Recreation in an Abandoned Rail Yard in Berlin. In: Kowarik I, Körner S (eds) Urban Wild Woodlands. Springer, Berlin Heidelberg, pp 287–299

Kowarik I, Schmidt E, Sigel B (1998) Naturschutz und Denkmalpflege. Wege zu einem Dialog im Garten. vdf Hochschulverlag, Zürich

Kowarik I, Körner S, Poggendorf L (2004) Südgelände: Vom Natur- zum Erlebnispark. Garten + Landschaft 114(2):24–27

Latz P (1987) Die Hafeninsel in Saarbrücken. Garten + Landschaft 97(11):42–48

Latz P (1999) Eine einfache Frage, keine einfache Antwort. DISP 138. Veröffentlichung des Instituts für Orts-, Regional- und Landesplanung der ETH Zürich 35(3):14–15

Latz P (1999a) Schöne Aussichten. Interview in Architektur & Wohnen 5:95–102

LeRoy LG (1978) Natur ausschalten, Natur einschalten. Klett-Cotta, Stuttgart

Lindner W (1926) Ingenieurwerk und Naturschutz. Hugo Bermühler, Berlin

Mattern H (1950) Über die Wohnlandschaft. In: Mattern H (ed) Die Wohnlandschaft. Hatje, Stuttgart, pp 7–24

Mattern H (1964) Gras darf nicht mehr wachsen. Ullstein, Berlin Frankfurt/M Wien

Milchert J (1996) Sprachlos und geschwätzig. Garten + Landschaft 106(7):15–16

Mörsch G (1998) Denkmalbegriff und Denkmalwerte. Weiterdenken nach Alois Riegl. In: Kowarik I, Schmidt E, Sigel B (eds) Naturschutz und Denkmalpflege. Wege zu einem Dialog im Garten. vdf Hochschulverlag, Zürich, pp 89–107

Politische Ökologie (1995) Bitte nicht berühren! Ist der Naturschutz museumsreif? Politische Ökologie 13(43)

Popper KR (1972) Naturgesetze und theoretische Systeme. In: Albert H (ed) Theorie und Realität. Ausgewählte Aufsätze zur Wissenschaftslehre der Sozialwissenschaften. Mohr, Tübingen, pp 43–58

Rebele F, Dettmar J (1996) Industriebrachen. Ökologie und Management. Ulmer, Stuttgart
Reichholf JH (1994) Kampf an den falschen Fronten. Die Zeit, 1.7.1994
Reichholf JH (1996) In dubio pro reo! Mehr Toleranz für fremde Arten. Nationalpark 91(2):21–26
Reicholf JH (1997) Sine ira et studio. Nationalpark 95(2):19–21
Riegl A (1903) Der moderne Denkmalkultus. Sein Wesen und seine Entstehung. Reprint in Riegl A (1929) Gesammelte Aufsätze. Dr. Benno Filser, Augsburg Wien, pp 144–193
Rink D (2005) Surrogate Nature or Wilderness? Social Perceptions and Notions of Nature in an Urban Context. In: Kowarik I, Körner S (eds) Urban Wild Woodlands. Springer, Berlin Heidelberg, pp 67–80
Rudorff E (1897) Heimatschutz. Nachdruck 1994. Reichl, St. Goar
Schemel HJ (1997) Naturerfahrungsräume – Flächenkategorie für die freie Erholung in Naturlandschaften. Natur + Landschaft 72(2):85–91
Scherzinger W (1990) Das Dynamik-Konzept im flächenhaften Naturschutz, Zieldiskussion am Beispiel der Nationalpark-Idee. Natur und Landschaft 65(6):292–298
Scherzinger W (1991) Biotop-Pflege oder Sukzession. Garten und Landschaft 101(2):24–28
Scherzinger W (1995) Blickfang - Mitesser - Störenfriede. Nationalpark 88(3):52–56
Scherzinger W (1996) Naturschutz im Wald: Qualitätsziele einer dynamischen Waldentwicklung. Ulmer, Stuttgart
Scherzinger W (1997) Tun oder unterlassen? Aspekte des Prozessschutzes und Bedeutung des "Nichts-Tuns" im Naturschutz. In: Wildnis - ein neues Leitbild!? Möglichkeiten und Grenzen ungestörter Naturentwicklung in Mitteleuropa. Berichte der Bayerischen Akademie für Naturschutz und Landschaftspflege, Laufen/Salzach 1:31–44
Schoenichen, W (1942) Naturschutz als völkische und internationale Kulturaufgabe. Eine Übersicht über die allgemeinen, die geologischen, botanischen, zoologischen und anthropologischen Probleme des heimatlichen wie des Weltnaturschutzes. Fischer, Jena
Schwarz U (1980) Der Naturgarten. Krüger, Frankfurt/M
Seifert A (1941) Im Zeitalter des Lebendigen. Müllersche Verlagshandlung, Dresden/Planegg
Stoll S (1999) Akzeptanzprobleme bei der Ausweisung von Großschutzgebieten. Ursachenanalyse und Ansätze zu Handlungsstrategien. Peter Lang, Frankfurt/M
Sukopp H, Kowarik I (1988) Stadt als Lebensraum für Pflanzen, Tiere und Menschen. Forderungen an die Stadtgestaltung aus ökologischer Sicht. In: Winter J, Mack J (eds) Herausforderung Stadt. Aspekte einer Humanökologie. Ullstein, Frankfurt/M, pp 29–55
Trepl L (1992) Stadt-Natur. Ökologie, Hermeneutik und Politik. Rundgespräche der Kommission für Ökologie, Bd. 4. ‚Stadtökologie'. Pfeil, München,pp 53–58

Trommer G (1992) Wildnis – die pädagogische Herausforderung. Deutscher Studienverlag, Weinheim
von Salisch H (1911) Forstästhetik. Dritte vermehrte Auflage. Julius Springer, Berlin
Wächter M (2003) Die Stadt: umweltbelastendes System oder wertvoller Lebensraum? Zur Geschichte, Theorie und Praxis stadtökologischer Forschung in Deutschland. UFZ-Bericht Nr. 9, Leipzig-Halle
Wiersbinski N, Erdmann K-H, Lange H (eds) (1998) Zur gesellschaftlichen Akzeptanz von Naturschutzmassnahmen. BfN-Skripten 2. Bundesamt für Naturschutz, Bonn

Approaches for Developing Urban Forests from the Cultural Context of Landscapes in Japan

Ryohei Ono

Graduate School of Agricultural and Life Sciences, University of Tokyo

Introduction

As it has often been said, we need to consider the delicate balance between human activities and the natural environment for the sustainable development of our living world. What seems to be important here is that in discussing this balance, we need to take into consideration the cultural aspects of the relationship between humans and nature as well as the physical aspects.

Each city stands in a place (see e.g., Relph 1976; Tuan 1977) with its unique historical background, and it is irreplaceable. However, after the Industrial Revolution, some cities were developed as "industrial facilities", their only function being production and distribution. These cities could stand wherever functional spaces (not places) were available, and they didn't need the support of accumulated layers of history.

Now in this post-industrial age, the problems of handling these kinds of cities are very challenging. Actually, converting these industrial spaces, so-called brownfields, into green fields such as forests may be viable. However, it seems to us that the problems cannot be solved by anti-industrial thinking such as "back to nature" or "ecological city" unless we also consider the historical or cultural characteristics of the places. The aim of this paper is to show how such cultural aspects can be included in the discussion about the problems of forests in cities. In other words, how can these new types of forests become the new landscape woven into the fabric of cultural context (Spirn 1998)?

These problems are really universal issues all over the world, so it is all the more necessary to consider the diversity of each culture. For instance, in Japan, as a part of the Asian region, many communities in the villages, towns, or cities have coexisted with forests for a long time, unlike in typical and traditional European cities. In Japan, there are many traditional

Kowarik I, Körner S (eds) Wild Urban Woodlands.

woodlands (nature of the first or second kind as defined by Kowarik 2005) within cities, as mosaics of green. The aim of this paper is to review the historical background of this coexistence in Japan, and to discuss some problems regarding managing forests in the future, from the viewpoint of the cultural context of landscapes. It is, of course, true that the Japanese have not always been kept lush forests across the whole of Japan. There have been eras of exploitation or destruction of forests for timber or firewood in Japan also (Totman 1989). However, these cases were mostly in the peri- or non-urban woodlands, and the scope of this paper is the cultural relationships between city people and the urban woodlands within cities.

Forests and religious belief

It is believed that the major factor in the coexistence of forests and cities in Japan is the religious belief called Shinto (see e.g. Ono 1962; Pregill and Volkman 1999; Sonoda 2000). The distinctive feature of this Japanese primitive religious belief is nature worship. In ancient times, various natural objects or places were regarded as sacred, and still today we can see their relics easily (Figs. 1, 2). In Japan, people tend to believe that these places are possessed by invisible presences such as native spirits.

Fig. 1. (left) An example of worshipping a big stone (in Yamanashi-shi, Yamanashi, 2003/12/25)

Fig. 2. (right) An example of worshipping a hill (in Kasugai-cho, Yamanashi, 2003/12/25)

Especially in Japan where the land is mostly mountainous, "forest" was synonymous with "hill" or "mountain", and they were regarded together as a sacred area. Sometimes the sea was also regarded as sacred. After people started to grow rice on irrigated flat fields, forests became indispensable as sources of water with the result that people strongly recognized forests as sacred, and built shrines at the boundary between forests and villages. Until the Middle Ages, many cities were developed in the limited flat land between the mountains or between the mountains and the sea. Inside these cities, major shrines were built on or at the foot of the small forested hills.

It is believed that for city people, the image of the hill composed of a forest and a shrine is analogous to the ancient setting of a forested mountain and a village's shrine built in front of it (Fig. 3).

After Buddhism was introduced into Japan from China in the sixth century, the fusion of Buddhism and Japanese Shinto became more and more advanced, and Buddhist temples and Shinto shrines came to coexist on the same site surrounded by forests.

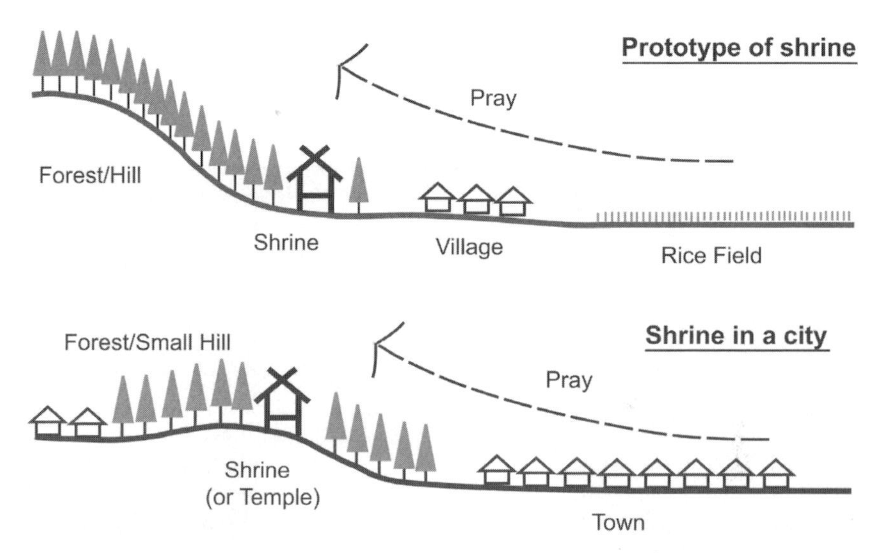

Fig. 3. Diagrams of spatial relationship between shrines and surroundings - prototype of shrine (above), shrine in a city (below)

Shrines and forests in cities in pre-modern times

From medieval times till the early modern times, the shrines or temples and their surrounding forests in the cities tended to become landmarks and core centres of the communities. The phenomenon was not unique to Japan; religious buildings all over the world shared this tendency as well.

What is remarkable, however, is that in Japan these places had roles as spaces for interactions between humans and nature—spaced known in modern terms as recreational areas or public parks (Conway 1991). People enjoyed the scenery or seasonal blossoms, and participated in festivals held within the grounds of the shrines or temples and their surroundings. Especially under the long and stable reign of the Tokugawa Era, i.e. from the 17th century to the middle of the 19th century, these activities extended beyond the religious grounds and into the entire city.

As these activities were linked to nature, the seasonal programmes were familiar to the people. Some kinds of illustrated programmes or pictorial guidebooks about these shrines or temples or other sights, were printed and published, and become popular among the people (Figs. 4, 5). Thus, these forested shrines or temples came to be regarded as noted attractive sights within the cities. What is important here is that the people had come to regard the shrines or temples and their surroundings with some kind of accumulated narratives (Lapka and Cudlinova 2003) beyond their pantheistic religious belief.

Fig. 4. A series of Yedo-meisho-zue (popular pictorial guidebooks of Tokyo published in 1836)

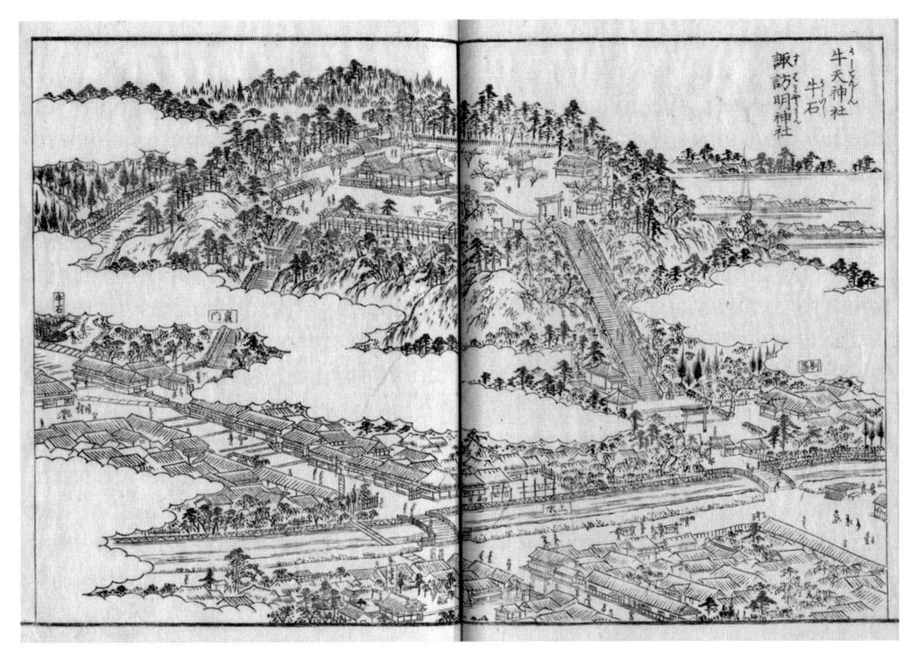

Fig. 5. An example of an illustration of a shrine and surrounding forest in Tokyo (from "Yedo-mesiho-Zue")

Shrines and forests in cities in modern times

After the middle of the 19th century, Japan abandoned its national isolation policy, and stepped drastically into the modern era under the strong influence of the Western world's advanced technologies. However, these were also times when nation states were being born in the West (Anderson 1983), and this influenced the many cultural or institutional conversions intended to enforce the integration of a national Japanese identity. The new Japanese government brought back the Tenno as a Westernised emperor, whose status once had been the emperor of Japan, and who then had been forced to the sidelines in the Tokugawa Era. The government also claimed its legitimacy in combination with the traditional Shinto religion. At this point, the pantheistic characteristics of the religion and the accumulated narratives were discarded, and only the visible Tenno family was regarded as existing and as descendants of the gods of the ancient Japanese myth. Consequently, all the shrines were registered in order of hierarchy, and large national shrines were respected and preserved, while small or primitive shrines were abolished or made to merge inside the municipal area.

On the other hand, some sites of shrines or temples were designated as public parks, and several sites were actually conserved as public open spaces. However, the major purpose of these changes was to support the national land-management policy, so there was little motive to conserve these places.

Early in the 20th century, the government planned a national project to build a shrine in Tokyo (Meiji Jingu) after the death of the reigning Tenno to worship him as a god. For the project, an artificial large (approximately 70 ha) forest was also to be laid out surrounding the shrine (as nature of the third kind; Kowarik 2005). These plans were realised, and today the forest is admired for its scientific vegetation planning and for the foresight involved in planting a forest in a megalopolis. However, unlike pre-modern times, the forest's characteristic as a place for interactions between city people and nature has almost disappeared. The narratives that had been shared among the people concerning shrines or temples and their forests were converted to the national myth to bless the Tenno, and the myth was misused for promoting extreme nationalism and fascism to head the nation toward World War II.

Shrines and forests in cities in post-war times

After World War II, Japan, as a defeated nation, was forced to reform its religious policy so as not to return to its former totalitarianism. The separation of government and religion was carried out, and that also had influences on the shrines or temples and their forests. In reality, all public parks that had been designated on the sites of shrine or temples were abolished. These places remained as spaces with woodlands, but these special areas where once people enjoyed nature and then believed in the distorted myth of the nation, became just disordered and untidy areas with lots of trees and religious buildings. In other words, the narrative of the relationship between city and nature, which had been shared among people in pre-modern times, was first carried in an eccentric way through the process of modernization, and then was forced to disappear as an extreme reaction.

As a result, the forests surrounding the shrines or temples tended to be destroyed and developed for their economic value. Many shrine or temple forests were converted to office buildings, commercial centres, or parking lots as real estate ventures (Fig. 6), while still retaining their religious function (Shelton 1999). Today the importance of these forests is being recognized again, but this is limited to an ecological interest in urban vegetation, and the social value is hardly recognized.

Fig. 6. An example of economic development on a site of a temple in Tokyo

Conclusions: Post-industrial cities and the possibility of new forests

Today in Japan, how to manage the landscape in this post-industrial age is also of public concern. One solution being considered is creating forests by planting trees on the brownfield sites where there once were industrial factories (nature of the fourth kind; Kowarik 2005). It is important, however, that we respect the cultural context concerning the relationship between cities and forests.

One aspect that should be considered is spatial axes and their directions in cities. In Japan, lands used for heavy industry have expanded the city towards the coastal area (Zukin 1991). Simply planting forests on these lands could reverse the direction of the cities' spatial axes from the traditional mountain-wards direction to the unknown sea-wards direction (Fig. 7). Whether the change is acceptable for the city is something that must be considered.

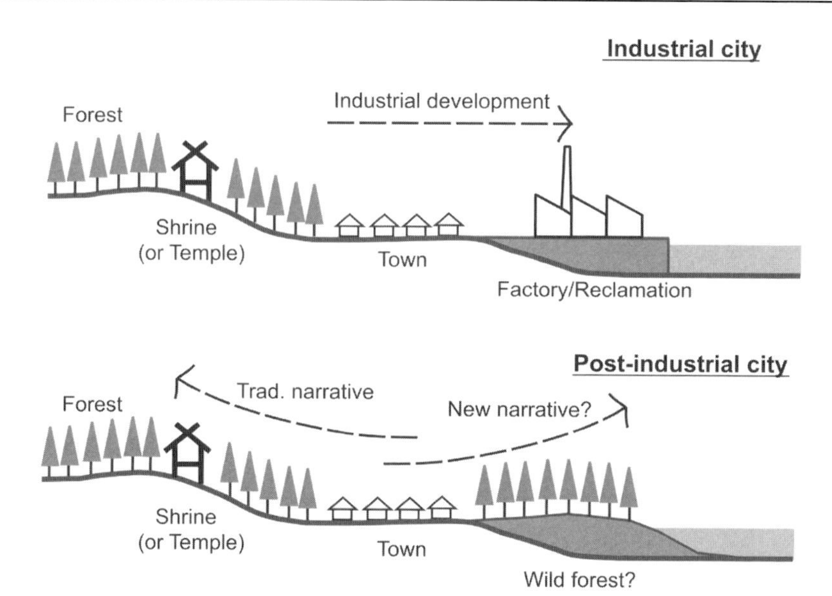

Fig. 7. Diagrams of industrial development (above) and the different narratives in the post-industrial city (below)

Furthermore, it can be said that, in Japan, the landscape of forests in the city was a great narrative itself—a bygone memory of an agricultural landscape, discovered by city people in the post-agricultural age. This being the case, we should discuss how to treat the historic industrial landscape in a way that is culturally acceptable to the people of the post-industrial age. We should consider forests of the post-industrial age in a different way from forests of the post-agricultural age (Fig. 8).

These viewpoints, which take into account the long history of civilization, seem meaningful to us not only in Japanese cases. Even if we could build forests in the cities using ecological technology without any cultural context, we cannot be sure that the forests would be accepted by people and built into the new narratives of the cities. On the other hand, however, we should also take notice of the both spontaneous and enforced characteristics of the narratives mentioned above. It is true that the narratives of landscapes are social and cultural products and ways of seeing (Cosgrove 1984). In any event, we must frequently remind ourselves that the problem of landscape in the post-industrial age should be fully discussed as an issue concerning the history of civilization and culture.

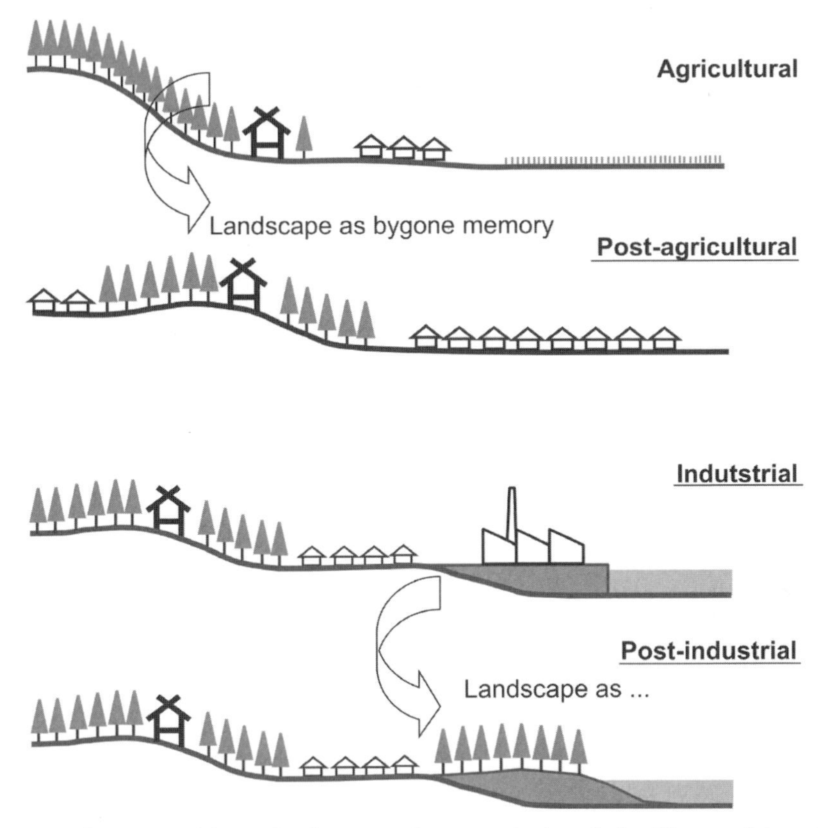

Fig. 8. Diagrams of forest landscapes of post-agricultural significance (above) and post-industrial significance (below)

References

Anderson B (1983) Imagined communities – Reflections on the origin and spread of nationalism. Verso, London

Conway H (1991) People's park – The design and development of Victorian parks in Britain. Cambridge University Press, Cambridge

Cosgrove DE (1984) Social formation and symbolic landscape. Croom Helm, London

Kowarik I (2005) Wild urban woodlands: Towards a conceptual framework. In: Kowarik I, Körner S (eds) Urban Wild Woodlands. Springer, Berlin Heidelberg, pp 1–32

Lapka M, Cudlinova E (2003) Changing landscapes, changing landscape's story. Landscape Research 28:323–328

Ono S (1962) Shinto – the Kami way. Tuttle, North Clarendon

Pregill P, Volkman N (1999) Landscapes in historyDesign and planning in the Eastern and Western traditions. 2nd edn John Wiley & Sons, New York
Relph E (1976) Place and Placelessness. Pion, London
Shelton B (1990) Learning from the Japanese city – West meets East in urban design. E & FN Spon, London
Sonoda M (2000) Shinto and the natural environment. In: Breen J, Teeuwen M (eds) Shinto in History – Ways of the Kami. University of Hawai'i Press, Honolulu, pp 32–46
Spirn AW (1998) The language of landscape. Yale University Press, New Haven
Totnam C (1989) The green archipelago: Forestry in pre-industrial Japan. University of California Press, Berkley
Tuan YF (1977) Space and place – The perspective of experience. University of Minneapolis, Minneapolis
Zukin S (1991) Landscapes of power. University of California Press, Berkley

Strategies between Intervening and Leaving Room

Lucia Grosse-Bächle

Institute of Open Space Development and Planning Related Sociology, University of Hannover

Designing within the dynamics of vegetation

"Wild Woodlands in the City" is not only a topic for ecologists and landscape planners. As a task involving design and communication, the theme challenges landscape architects to reflect on the aesthetic dimensions of the post-industrial urban landscape. Here, new concepts and models that integrate spontaneous nature and how these might look are discussed.

This paper draws attention to the natural dynamics of vegetation and how these can be used creatively. Possibilities drawn from contemporary landscape architecture illustrate ways to initiate, to work with or to use these dynamic processes in new ways.

The nature of the wild woodland

The term "wild woodland" is used in this context to describe freely developing, essentially undisturbed nature. Since the Age of Romanticism, wilderness and the forest have embodied the longing of humans for self-determination and for the development of our own inner spirit. In a society that was increasingly shaped by rationalism, the forest was seen as a place outside civilization. Here emotions could be expressed that otherwise would not be tolerated (Braun 2000: 51). Many people today, in a world primarily organized around performance and practicality, are looking for the experience of the "unattainable" (Seel 1996: 189). To allow confrontations with the impersonal power that is nature, nature should be allowed to develop according to its own rules wherever space allows.

Kowarik I, Körner S (eds) Wild Urban Woodlands.

Design as a form of intervention

Older still than the Romantic-era longing for untouched nature is the human need to transform the environment according to our desires. The physical structures that have been left behind as clues in the landscape reflect the values and principles of the societies from which they came. In the process of design, the designer intervenes in the natural environment and grapples with it intensively.

In the opinion of the anthropologist Arnold Gehlen, humans would be unable to survive in an entirely undesigned environment because the role that the environment plays in the world of animals is played in the human world by culture. Culture is the part of nature that has been overcome by humans (Franzen 2000).

Dialogue between humans and nature

Both the need for wild and undisturbed nature and the need to design nature and the environment are deeply rooted in humans and both must be valued. As the writings from 1559 of the Italian humanist Bonfadio show, the idea of a synthesis, or a partnered relationship between humans and nature is nothing new: "The garden is the product of an almost alchemistic development process in which nature and art are involved in equal parts and where natural artificiality and artificial nature alternate and combine with one another" (De Jong 1998: 22). Through the combination of both in the creative process something new is created, the so-called third nature.

In order that the idea of a dialogue between humans and nature be given appropriate expression today, a new, contemporary aesthetic language must be developed. This dialogic relationship is reflected most clearly in design with the living material plants.

A plant is not a stone

A prerequisite for innovative design with plants is the rediscovery of their specific qualities. With unprejudiced perception, free from preconceived ideas, one is in a position to recognize the exceptional and to pursue new avenues of development. A comparison of materials makes clear that plants differ from other media less through their outward form and far more through their exceptional potential for adaptation and their ability to direct themselves. Plants are both the active opponents as well as the passive medium of the landscape architect. This double nature confronts the designer with an unsolvable contradiction that can be seen as both a problem and an opportunity.

Landscape architects who wish to design with plants must deal with many different facets because of plants' complex nature. Not only aesthetic-cultural aspects play a role, but biological and practical issues must also be considered.

- If one considers the plant as a biological phenomenon, the most interesting aspect is the multitude of dynamic processes that can be seen at all levels of the plant's existence.
- If one considers the plant as a medium and therefore, as an aesthetic-cultural phenomenon, it conveys fascinating ideas, meanings and models for living.
- If one considers the adaptability of plants as a methodological challenge, it is interesting to consider the methods and strategies available for reacting flexibly to the unexpected.

Space for dynamic processes

The prerequisite and, at the same time, goal of designing with dynamic processes is a sensitivity to and a respect for the beauty of that which arises spontaneously (*Von-Selbst-Entstanden*). A well-informed and careful shaping of spaces to allow the self-development of vegetation is required. At the same time, a helping of courage and a playful openness is necessary because in this regard landscape architects put themselves at the helm of evolution. This is a task that must be combined with knowledge and responsibility lest one mutate into a “green Frankenstein.”

Landscape architects who design with vegetation find themselves in the role of “protectors of creation” as well as in the role of “creator” itself. The oscillation between the two ends of the spectrum “allowing room for self-regulating processes” and “intervening through design” is characteristic of work with living plants.

Idealization of natural images as an inhibition of creative design

Until now, the dynamic properties of plants have played a relatively subordinate role in the designs of landscape architecture. Toleration of dynamic processes in the history of gardens was usually linked with an image of intact and undisturbed nature. This idealized vision made it difficult for the designer to use the dynamics of plants creatively and to playfully integrate plants into designs. Process-oriented design seemed to be bound a priori to a semi-natural aesthetic and therefore allowed no designed interventions that could be recognized as manmade (Grosse-Bächle 2003: 116) Only a sufficiently open vision of nature makes creative handling of the dynamics of plants possible.

Current examples of contemporary landscape architecture show that a new willingness exists to be free from inhibited ideas about nature and to experiment freely with the dynamics of vegetation. Strategies and methods based on accepted responsibilities toward nature have been developed that allow vegetation room to develop freely while permitting interventions to guide the development process.

The following examples illustrate possibilities that exist for incorporating the natural dynamics of vegetation into design. The inherent dynamics of the natural development of vegetation are made perceptible and guided in different ways.

Design strategies

If the natural dynamics of vegetation are to be used in process-oriented design in order to create flexible and suitable spaces, a more exact knowledge of plant-specific strategies is needed. In population biology and plant sociology, the term "strategy" indicates the combination of genetically determined physiological and morphological adaptations for dominating a site under optimal use of resources. This includes processes from self-thinning to succession. In the following design examples individual elements of such processes are selected and manipulated through design.

Using design to work with processes: Oerliker Park

In their design for Oerliker Park, the planning group Zulauf, Seippel, Schweingruber, Hubacher & Haerle uses the ability of spaces to adapt to changing needs as their central idea. The park is intended to be realized in phases in order to accommodate the growth of "Central Zurich North", a new part of the city.

Because future parameters are uncertain, the planners designed an all-purpose framework - a kind of green building - that can be filled over time.

The designers took up the theme of change, and young ashes were planted in a dense grid, interspersed with fields of cherry, sweetgum and pawlonia trees. These will gradually grow into a bright "hall of trees". The development process will be guided by maintenance measures that function as designed interventions and that bring a unique quality to each phase of the development (Figs. 1–4).

Fig. 1. Planting the ashes in Oerliker Park in 2001, simple saplings/ forest nursery products

Fig. 2. One year later, sizeable trees have grown from the meagre saplings (photo: Volkmar Seyfang)

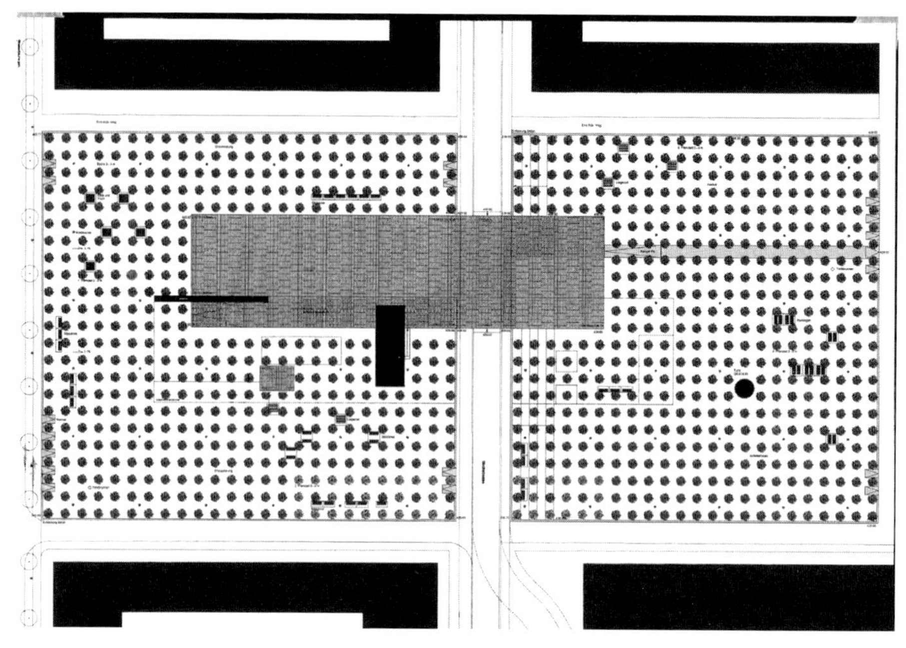

Fig. 3. Design for Oerliker Park. Initial stage: a grid of young trees defines the space

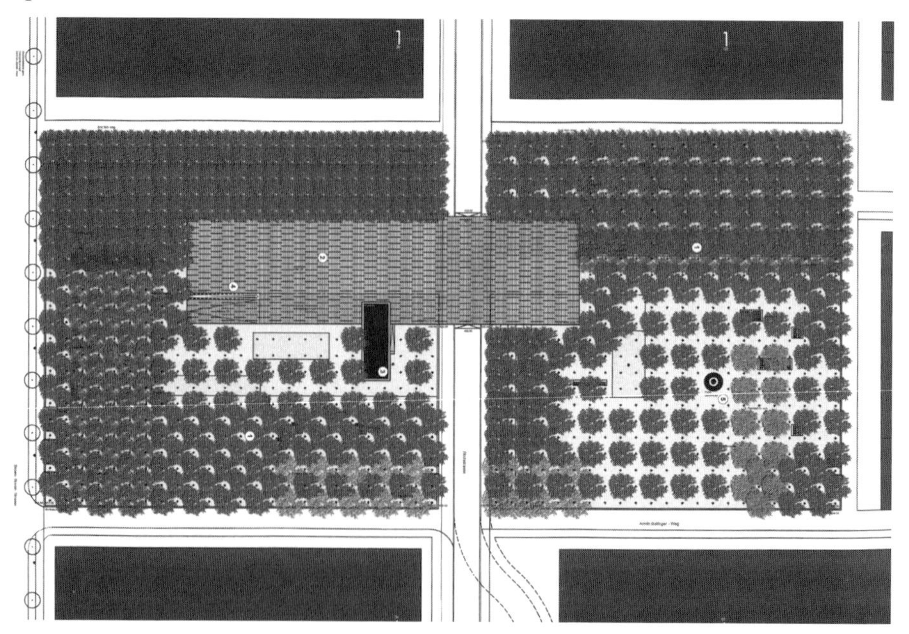

Fig. 4. Design for Oerliker Park. End stage: The structure of the space is determined by changing the spacing of the tree grid (reproduced with permission by Zulauf, Seippel, Schweingruber).

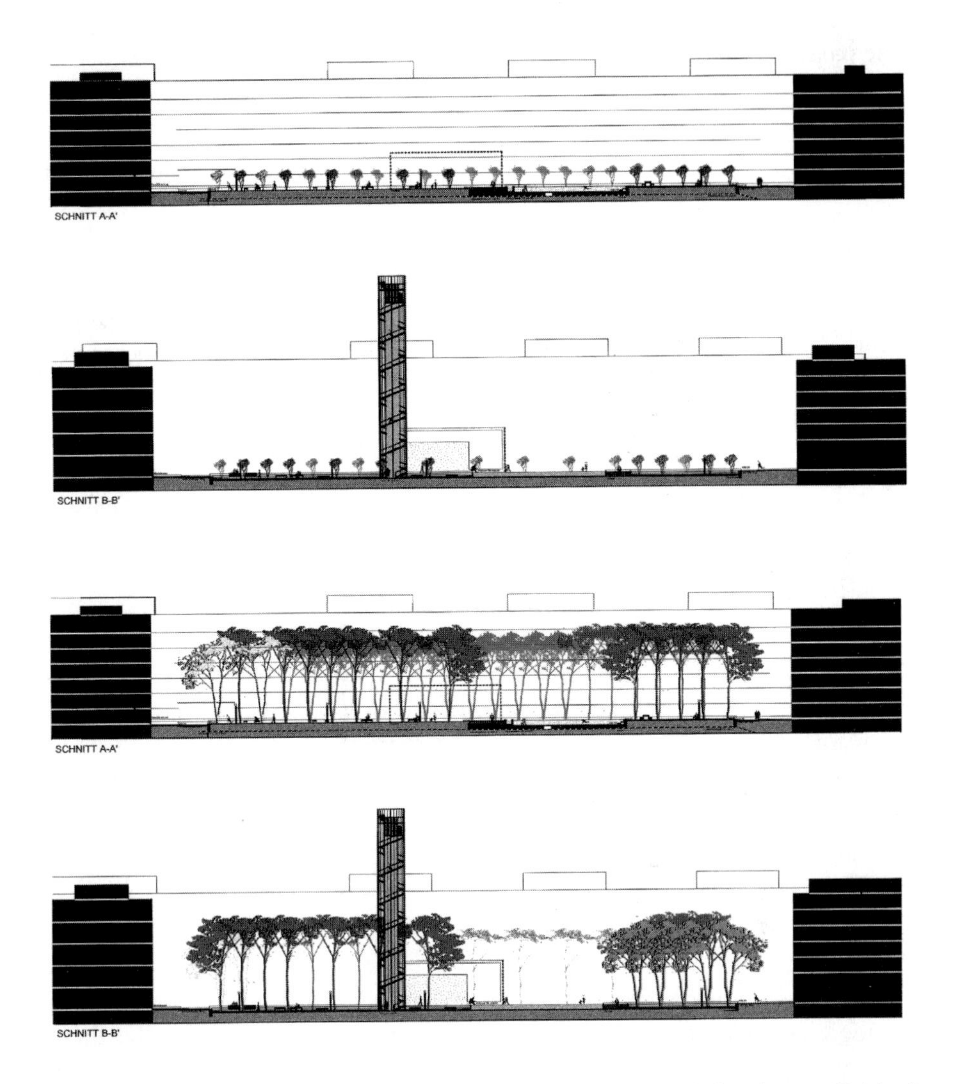

Fig. 5. Sections through the park show the development of the dense "field of trees" to the spacious "hall of trees" (reproduced with permission by Zulauf, Seippel, Schweingruber)

A detailed development plan lays out the rules for guiding the intervention. Over decades, the dense tree grid of 4x4 m^2 will gradually be widened to 8x8 m^2 in places. The landscape architects have appropriated the thinning process from forestry; it is comparable to thinning a forest stand. Trees that die will not be replaced. In this way, the grid will slowly dissolve. From an initial, stark regularity, a certain irregularity will develop as time passes (Fig. 5).

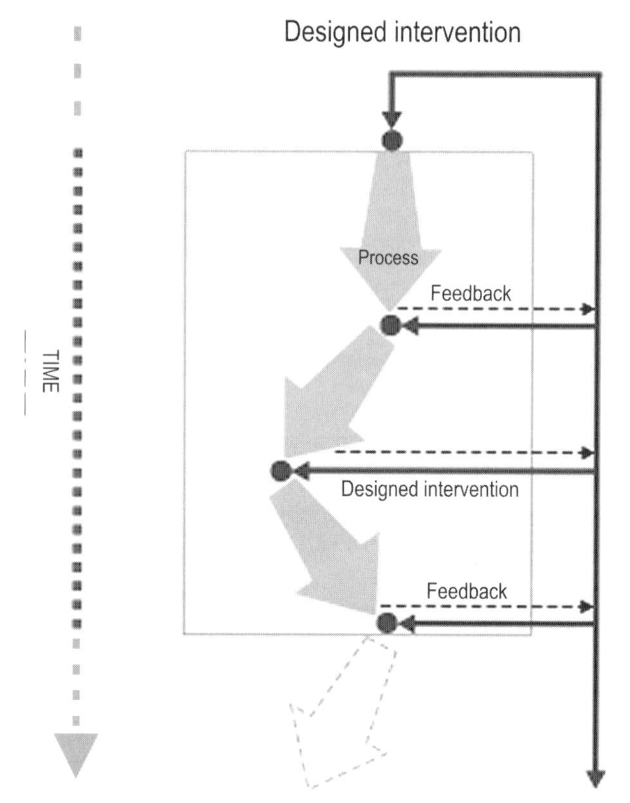

Fig. 6. Guiding the development process at Oerliker Park: With the planting of young ashes, the landscape architects direct the transformation and growth process. The process is guided through maintenance interventions based on feedback, i.e. after assessing the existing situation (after Grosse-Bächle 2003).

The vegetation concept for Oerliker Park is principally determined by strategic considerations. The landscape architects have grasped the theme of ever-present change and playfully influenced the growth process of the young ashes. They make creative use of the survival strategies of wild plant populations, such as the process of self-thinning, to suit the changing needs of the space.

Planting trees in an open space at regular intervals is comparable to the initial set-up of a game (Schmid 2001). As in a board game (e.g. Go), new configurations arise over time through planned interventions. The rules of the game are familiar, the final positions only approximately predictable. A differentiated structure of the space, marked by dense and open spaces will develop step by step through the thinning of the homogeneous field of trees. A shady passage, a bright clearing, an open space will have replaced the regular grid in 25 years. By highlighting and giving value to intermediate stages, these stages take on a new quality (Fig. 3–4).

More a manifesto than a plan: the grounds of the Millennium Dome

The French landscape architects Desvigne & Dalnoky make use of a similar method in their concept for the Greenwich Peninsula Park. They consider the permanence and the transformability of the landscape to be solid principles from which to react to the imponderability of the urban development. In this way, the concept for the London park seems more a manifesto than a classic design plan (Arnold 1998).

According to the project requirements, the planning for the outside spaces was to operate along two time lines: in the short-term, for the open spaces associated with the millennium celebrations and in the long-term, for the development of a new part of the city. A landscape was planned around the Millennium Dome that would develop in phases. As an initial spark, generous poplar plantings were specified, overlaid with a grid of individual trees spaced 3.5 m apart. The grid is to be thinned step by step to a spacing of 7x7 m^2 until eventually the planting consists only of groups of trees. Desvigne & Dalnoky describe a "strategy of invasion", as they structure the entire area around the Dome with young, fast-growing poplars planted close together. In a manner analogous to the processes of natural succession, the landscape architects combine pioneer species with climax species. At the beginning of the process, the plentiful pioneer plants will dominate the park; in the following phases, the climax species with their powerful form will determine the vision. Out of a dense, forest-like planting, an interesting landscape organized around clearings and paths will gradually emerge.

A landscape for the future: the Thomsen plant in Guyancourt

In a very similar way, Desvigne & Dalnoky have created a concept for the Thomsen plant in Guyancourt near Paris. The project required them to prepare an inhospitable landscape for the construction of a factory and a parking lot for a thousand cars in a very short time at minimal cost. Factories generally have a limited life span, no more than decades – the span of time a garden needs to fully ripen. The landscape architects envision that the vegetation around the buildings will have developed fully by the time the buildings are ready to be torn down. “Our desire is to plan all stages of development and to allow them to be experienced during the lifespan of the industrial complex and beyond” (Desvigne and Dalnoky 1994: 22).

For the first stage of the factory site, drainage ditches are laid and planted with willow and poplar groups. Because these pioneer species grow quickly, a dense vegetation develops rapidly on the site: the first evidence of a new landscape. In later stages, saplings of large trees will be planted in the poplar groups. During the first 15 years of their growth, the little plants will hardly be noticeable; later they will overtake the poplar and willow pioneers. In addition, plantings of black pine and about 100 free-standing conifers will be added. While the fast-growing poplars quickly envelop the factory buildings, the slow-growing pines come to define the character of the landscape much later. “Only hints of the first planting will be recognizable when a park with choice plantings emerges at the time the factory buildings are demolished” (Desvigne and Dalnoky 1994: 23).

Strategies of the “green guerrilla”: the Schipol airport

On the expansion site of the Schipol Airport in Amsterdam, the landscape architects of the planning firm West 8 have planted 800,000 birches (Fig. 7). They chose the plants based on ecological expert opinion from the state forestry institute because birches are particularly suitable for the greening of airport grounds. None of the birds of prey that disturb the workings of airports can nest in their soft twigs. Birches, as r-strategists, belong to the group of fast-growing pioneer species with good dispersion potential in growing sites without much established vegetation. In the initial phases of establishment, a large number of individuals are dispersed that are reduced in later phases through various means such as competition or disease. The landscape architects made use of these natural survival strategies of the pioneer species, and for six years, without a particular design, they planted every conceivable bit of land: medians, cable routes, inner courtyards of the parking lots. “We were a kind of green guerilla and people fought against us everywhere where the trees were supposedly bothersome” (Geuze quoted in Weilacher 1996: 236). An underplanting of clover was planned to improve the growth of the trees.

Fig. 7. The Schipol airport planting plan: strategies of the "green guerrilla"

For Adrian Geuze of West 8, designing landscapes always includes organizing processes. It isn't sufficient to undertake a few designed interventions -- all individual components must be a part of a whole concept, a strategy. A concept for long-term development of the landscape takes time. Twenty years might pass from the beginning of the process until all the individual steps have unfolded completely.

This method of working allows Adrian Geuze to design through a dialogue with nature: Through the interplay between action and reaction, between design and natural process, Geuze attempts to influence the laws of nature through playful means. He firmly believes in the success of evolution, in the "survival of nature".

"I would like to bring people back in touch with their contemporary landscape. I don't want to create new illusions that reinforce the preconception that our landscape is ruined, that our society is bad, that we destroy the landscape, that we manipulate the entire planet, that we will soon be dead, we must protect ourselves, and enshrine our landscape. I don't care for this pessimism" (Geuze quoted in Weilacher 1996: 234).

Fig. 8. Natur-Park Schöneberger Südgelände: A path connects the pieces together

A mosaic of different dynamics: the Natur-Park Schöneberger Südgelände

In the near future, ongoing succession would have led to the complete reforestation of the Schöneberger Südgelände (Fig. 8). Instead, the decision was made to maintain the rich diversity of form through maintenance and development measures. In certain areas, a few "ruderal woodlands" are to be given over entirely to succession to represent "urban wilderness".

With various maintenance interventions, the planners of the group Öko-Con and Planland have created different stages of succession in neighboring parts of the site (see also Kowarik and Langer 2005). Phases of succession that normally would be experienced over the course of time, can be experienced here as one moves from one space to another. A path leads the visitor through the heterogeneous spaces of the nature park. In this way, as one moves through the space the mosaic of succession stages is experienced as a time sequence (Fig. 9).

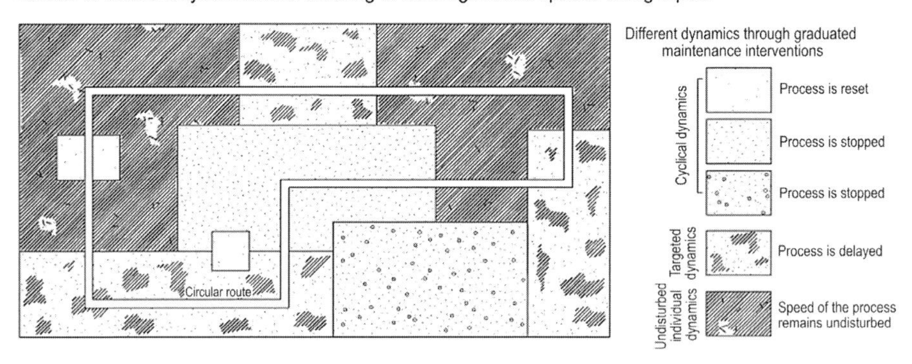

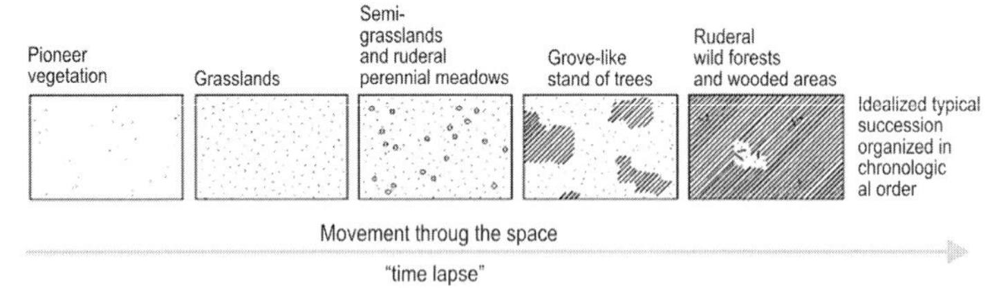

Fig. 9. Stages of succession, maintained by various interventions, are created in neighbouring spaces. The phases of succession that normally would be experienced over the course of time, can be experienced here as one moves from one space to another. As one moves through the space the mosaic of succession stages is experienced as a time sequence (after Grosse-Bächle 2003).

The natural dynamics of the Südgelände are slowed, the succession delayed.

- In the area of the grasslands and tall shrubs, time seems to be stopped by a cyclical recreation of a particular state.
- In the grove-like stands of trees, time is slowed, succession is directed.
- In the wooded areas, the rate of the natural processes is undisturbed and corresponds to the inherent dynamics of the stand (Fig. 9).

The spatial qualities of the site are highlighted with design interventions that serve to illustrate the value of the site. Elements of the infrastructure, such as industrial relics, viewing points and rest areas are concentrated along an iron "boardwalk", that only comes into contact with sensitive vegetation in a few spots. By bundling the functions in a single setting, large areas are kept free from intervention and space for undisturbed development is created (Fig. 10).

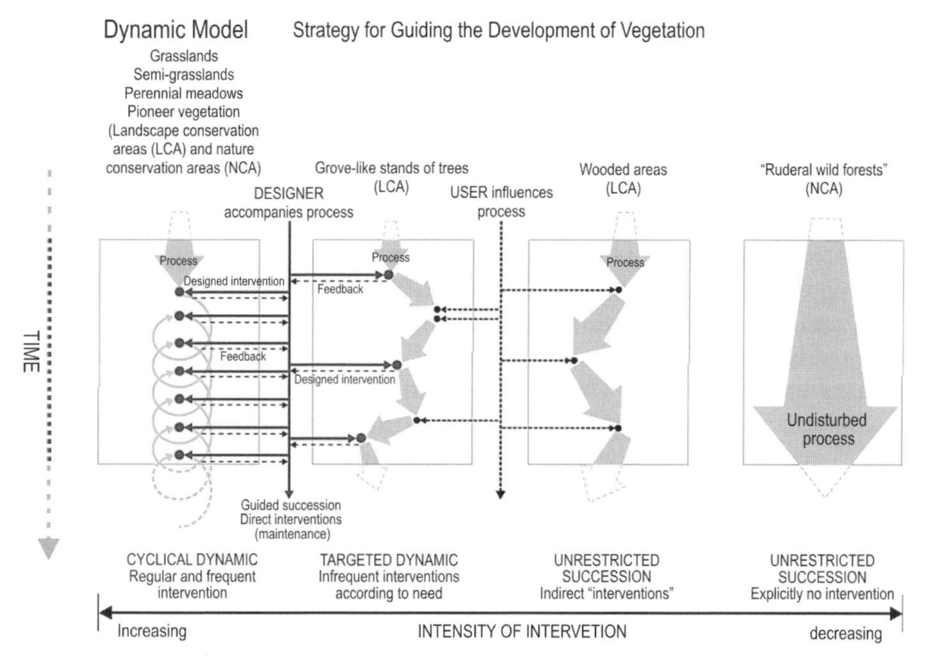

Fig. 10. Bringing together natural processes and guiding interventions: In neighboring spaces, the differing intensity of maintenance measures creates a cyclical dynamic, a directed dynamic or a dynamic indirectly influenced through use. In the "ruderal wild woodland", the process continues undisturbed (after Grosse-Bächle 2003).

A new vision for the urban landscapes of regions in decline

The search for new strategies for designing with the dynamics of plants will acquire greater meaning in the future. One of the most important tasks of landscape architecture will be the development of new visions for the urban landscape of declining regions. In this context the question of how to create suitable, cost effective and aesthetically valuable designs using vegetation will acquire increasing relevance. For many wastelands, particularly coal and steel industrial sites, the most ecologically sensible and aesthetically appealing solution is to allow vegetation to grow on its own. For financial reasons, this is also often the only possible way.

A *lot* of woodland, but not *only* woodland

Reforestation measures, planting for nature conservation, as well as spontaneous processes of succession in declining regions such as the Ruhrgebiet are leading to increasing forestation of developed areas. Even if the increase of wild woodlands in urban-industrial areas is unquestionably valuable, they do not have to be protected at any cost, their natural development needs not stay free of designing intervention. On the contrary, the denseness and closed-ness of spontaneous woodlands require design interventions in the places where the woodlands meet developed areas in order to open up the space, create lines of sight, and allow public use.

In addition to semi-natural cultivation, the development of artificial visions of the woodland, especially along the borders of developed areas, increases in significance. The woodlands no longer stand in opposition to developed areas, but rather are a part of the urban-industrial landscape. The “natural woodland” gives way to more transparent forms such as tree plantations, extensively maintained park landscapes, or agricultural land. A mosaic of open spaces and woodland types requiring differing intensities of care from wild industrial woodlands to production tree plantations open for recreation activities is possible by combining types.

Summary of the important characteristics of integrated design principles

The design examples described here show a new readiness to experiment with the dynamics of vegetation. The designers look for strategies for integrating, working with and manipulating the development processes of plants. Their goal is not only to protect nature, but to enrich it through ecological understanding. At the same time, they create space for self-organized processes and they respect the inherent dynamics of the vegetation.

The relationship between humans and plants is understood as an interactive process. To some extent, the specific methods of design and management resemble the work of directors. They don't assume a static picture, but rather a never-ending series of changing images.

The designers break away from conventional ideas about "beautiful nature" and experiment with the aesthetics of the everyday or the ugly. The states of growth, evolution and decay, i.e. that which is "unfinished", are valued as ecologically relevant. This philosophy is illustrated in the example of Oerliker Park in the integration of development stages of young plants and in the Südgelände nature park where dead trees are allowed to accumulate in the ruderal wild woodlandss. A positive vision of uncertainty is characteristic of the design principles illustrated here. Uncertainty is seen as a challenge, as an adventure and as room to maneuver. Changes aren't considered problems, but rather as enrichment and as the mark of successful life.

The examples above illustrate clearly that an incisive design vocabulary can be combined with design philosophy oriented toward natural processes. Planning that keeps natural processes sensitively in mind must not only serve a "semi-natural" or "imitation nature" aesthetic. An adaptable understanding of nature that starts by giving up the idea that man-made and natural are opposites and accepting that humans bear the responsibility for nature allows us to look forward to a new and unbiased future in our work with plants.

References

Arnold F (1998) A landscape for the Millenium Dome, London. Topos 23:98–101

Braun A (2000) Wahrnehmung von Wald und Natur. Leske und Budrich, Opladen

De Jong E (1998) Der Garten als dritte Natur. In: Kowarik I, Schmidt E, Sigel B (eds) Naturschutz und Denkmalpflege. Wege zu einem Dialog im Garten. vdf Hochschulverlag, Zürich, pp 17–27

Desvigne M and Dalnoky C (1994) A new landscape for the Thomsen plant in Guyancourt. Topos 7:20–24

Franzen B (2000) Die vierte Natur. Kunstwissenschaftliche Bibliothek Band 11. Verlag der Buchhandlung Walther König, Köln.

Grosse-Bächle L (2003) Eine Pflanze ist kein Stein. Beiträge zur räumlichen Planung 72. Inst. für Freiraumentwicklung und Planungsbezogene Soziologie, Hannover

Kowarik I, Langer A (2005) Natur-Park Südgelände: Linking Conservation and Recreation in an Abandoned Rail Yard in Berlin. In: Kowarik I, Körner S (eds) Urban Wild Woodlands. Springer, Berlin Heidelberg, pp 287–299

Schmid A (2001) Zwischen Kontrolle und laisser faire. Anthos 2:9–13

Seel M (1996) Eine Ästhetik der Natur. Suhrkamp, Frankfurt

Weilacher U (1996) Zwischen Landschaftsarchitektur und Land Art. Birkhäuser Verlag, Basel Berlin Boston

"New Wilderness" as an Element of the Peri-Urban Landscape

Sigurd Karl Henne

mühlinghaus + henne, BfL Mühlinghaus Planungsgesellschaft mbH

Introduction

Before the industrial revolution, landscapes changes were often not noticed because of their slow rate. Since then the increasing speed of these changes and their scale have made them visible.

As a result of urbanisation of the landscape near larger cities, a lot of farmland has been changed into traffic networks, housing and production areas (Losch 1999). The dramatic loss of agricultural elements has also changed the visual impression of the peri-urban landscapes in central Europe. Step by step they are losing their rural character.

Fig. 1. New areas with wild vegetation in the peri-urban landscape: long-term fallow field near Karlsruhe, southwestern Germany

Kowarik I, Körner S (eds) Wild Urban Woodlands.

At the same time, sites with wild vegetation become part of the peri-urban landscape. The growing number of these areas seems to indicate that wilderness is recapturing the landscape. But this *new* wilderness is different from the original wilderness that disappeared from central Europe centuries ago.

The following chapter tries to demonstrate that areas with spontaneous vegetation do not show only short-term signs of change. This new wilderness can become an important part of the peri-urban landscape in the future if concepts for its function and design are developed.

Potential areas of new wilderness in the peri-urban landscape

This section shows that in addition to areas that already have wild vegetation, there are different areas in the peri-urban landscape that could be potential places for new wilderness areas.

New wilderness on formerly cultivated land

As the landscape changes from a rural to a peri-urban one, potential areas for new wilderness are provided mainly by former farmland. One reason for this change is the development of different intensities of agriculture. With decreasing subsidies and strong competition in the European market, only land suitable for agricultural production can be profitable. Poorer land will either turn fallow or will be used for other urban purposes such as recreation (Losch 1999). In addition, better job opportunities in urban regions and the permanent loss of land due to urban development make profitable agriculture uncertain. For these reasons, farmland with very intense agricultural use can be found next to extensively cultivated or even fallow land. Although these areas appear wild they are not original wilderness, but new wilderness.

New wilderness on urban and industrial brownfields

The dynamic effects of globalisation cause constant changes of peri-urban land use and produce industrial and urban brownfields which are also potential places for a new, secondary type of wilderness. Today brownfields on former sites of heavy industry are well known for wild vegetation. The majority of future brownfields in urban regions will be found on small-scale commercial sites and only a few of them will be changed into public parks. The majority will have wild vegetation that is managed in a specific way to enable the reuse of the land in the future.

New wilderness for flora, fauna and habitats

Other areas for wild vegetation are created by legal regulations to compensate for the ecological losses of urbanisation. As a result, areas of new wilderness are established for flora, fauna and habitat purposes. A better understanding of the ecological benefits of dynamic changes to habitats has led to a change in strategy for the vegetation management of these areas (Wulf 1995) that aims at integrating natural development.

New wilderness as land reserve

New wilderness could also be situated on land that is temporarily not used, such as areas kept in reserve for urban development (especially traffic infrastructure). Some of these sites remain unused for longer periods because changes in planning and budget problems cause delays. The natural vegetation that develops on there could be managed as a (secondary) new wilderness as an interim solution.

New wilderness for land with extensive use

Extensively used public land is another area for wild vegetation. The vast "green" areas of traffic networks or commercial parks in the peri-urban landscape are examples of this type of open space. Today most of this land is cared for by traditional horticultural means. However, budget problems and ecological needs are leading step by step to new maintenance methods that integrate natural vegetation development in their strategies of extensive care, leading to vegetation forms that appear wild.

New wilderness as a method for conservation of rural landscapes

Today most strategies for the conservation of landscapes try to preserve rural character using traditional agricultural methods. However there is discussion about the aims and effectiveness of these methods. As a result of cuts in subsidies, these strategies have become questionable. Some methods that have been developed recently work with a strategy of minimal care, including the management of natural vegetation development. This could be another option for new wilderness in the peri-urban landscape.

Wilderness as an element of peri-urban park systems

As the peri-urban landscape becomes a functional part of the urban region its importance for recreational activities grows. New peri-urban park systems have been established recently on the periphery of cities (e.g. a park system northeast of Berlin and the Rhein-Main regional park near Frankfurt) that offer a new type of open space for recreation. The planners aim at a new identity for the peri-urban landscape by integrating different new elements. Due to the small amount of public funds and the large dimensions of these parks, it is very probable that areas with wild vegetation will be parts of these peri-urban park systems.

New wilderness for "shrinking" cities

New wilderness on unused land is not only found outside built-up areas. The present process of "shrinking" of cities in central Europe, especially in eastern Germany, provides sites that are no longer needed for further urban development. Nevertheless, concepts that provide urban qualities have to be developed for these areas. The need for maintenance and the small budgets available will probably also lead to methods of care that integrate the natural development of vegetation. The character of this vegetation cannot be really "wild", but its management should create a designed wilderness that helps prevent vandalism and littering. This design must correspond to aesthetic ideas of wilderness.

Character of new wilderness in the peri-urban landscape

The character of new wilderness is important for its acceptance by the public as an element of the peri-urban landscape. This chapter describes different characteristics of new wilderness areas that result from the conditions and functions of this landscape.

Wilderness – only one element of the peri-urban landscape

The constantly changing use of the peri-urban landscape is primarily influenced by the economics of land use. As new wilderness provides only little (or no) profit, it will probably cover only minor parts of the peri-urban landscape and will be only one element among others in the cultivated landscape. Although this seems trivial, it has important consequences for the perception of new wilderness as a part of the cultivated peri-urban landscape.

New wilderness as a dynamic secondary wilderness

The character of new wilderness in the peri-urban landscape is different from the original wilderness of nature of former times. As a product of functional changes, it is a secondary wilderness that shows remnants and traces of former cultivation. It consists of a variety of species and has a specific structure. As it is also a result of continuous internal changes, its appearance and structure are not stable, especially at the beginning.

Different types of wilderness

The new wilderness of peri-urban landscapes will display distinctive qualities in its different types of vegetation and structure that not only reflect its former use, but that relate to its contemporary function. The new wilderness will range from dense woodlands to pastoral landscapes with "wild" character.

The appearance of wildness through vegetation management

Although new wilderness vegetation in the peri-urban landscape will seem "wild", in most cases it will not grow uncontrolled but will be a result of vegetation management. Today, vegetation management is focused on developing habitats for ecological goals. In the future it will be even more specific as it will be aimed at various purposes. Therefore it will produce different types of vegetation. Most concepts for vegetation management will also have to find a way to provide small profits to finance the measures. Several types of extensive use (e.g. extensive pastures) will be necessary for this goal, which will also lead to different types of new wilderness in the landscape.

New wilderness to change urban brownfields

The long-term management of naturally developing vegetation can also help urban or industrial brownfields sites to be reused. However vegetation management must prepare these areas for new purposes in the future. As most of these purposes cannot be foreseen, one basic goal of vegetation management should preventing the growth of woodlands as this is an obstacle for further reuse.

At the same time, vegetation management can help to change the negative image of brownfields by developing stable and impressive vegetation structures. To reach this goal, vegetation management must also integrate design concepts (see below).

Design goals for vegetation management of new wilderness

Today most concepts for management of spontaneous vegetation do not include design goals. The appearance of the vegetation is just a side effect of the functional goals and conditions of management. However, aesthetic aspects are important for public acceptance of spontaneous vegetation (Keil 2002).

For this reason the following section will discuss the perception of new wilderness as a basis for design strategies.

Perception and acceptance of new wilderness in the peri-urban landscape

New wilderness – an aesthetic myth

The meaning of new wilderness plays an important role in its perception and acceptance as an element of the peri-urban landscape.

The understanding of new wilderness is related to the cultural concept of wilderness which must be seen as different from original wilderness. This new wilderness is also not the wilderness of traditional nature conservation, which describes a landscape without human interventions (see also the definition by Körner (2005).

The cultural concept of wilderness is based on a visual perception. It is an aesthetic perspective that can be seen as a compensation for the modern scientific perspective on nature (Grossklaus 1993). The cultural concept of wilderness does not create wilderness but images of *wildness* related to aesthetic and symbolic ideas of (wild) nature. The artificial wilderness of a holiday park also shows the relationship of image and meaning of wildness. The wildness of these parks is not the true, original wilderness but an image that has become reality through its relationship to positive understandings of wildness.

For this reason, the cultural concept of wildness is not defined by the degree of human impact but by an area's *wild appearance.* Wild vegetation does not have to be original nature but must show a wild character (Seel 1991). Even a secondary wilderness (of spontaneous vegetation) can be seen as expressing wildness, as long as it *seems* to grow uncontrolled (no matter if it is controlled by vegetation management).

The idea of wildness is important for the acceptance of wild vegetation in the peri-urban landscape. Its meaning today is the result of a process of changes. The understanding of wildness has changed from a negative symbol of a "place of horror" to a positive image of the self-determination of man and the autonomous complexity of nature. This change began even before Rousseau's concept of the natural freedom of man to which wildness is a corresponding image. Since that time, wildness has become an aesthetic myth of the modern society that incorporates different positive meanings.

The development of a positive meaning of wildness is the basis of an aesthetic perception of wild vegetation.

The concept of wilderness as the opposite of culture

Contrast is the most important way to support the visual and aesthetic perception of complex structures of wild vegetation. (Therefore its also *the* basic method for its design). The contrast between wild natural vegetation and regular man-made patterns also corresponds to a common cultural concept of wildness: the symbolic contrast between nature and culture (Bredekamp and Janzer 1985; Seel 1991).

The arts, literature and horticulture have formed images that describe this idea of wildness as the opposite of culture (or society, Grossklaus 1993). The symbolic contrast between wild nature and culture is the basis for the perception of wildness in the peri-urban landscape.

Industrial nature as an aesthetic concept for wildness

The IBA Emscher Park project has developed a new concept of *industrial nature* (Dettmar and Ganser 1999). This concept refers to a symbolic meaning of wildness: the natural development of vegetation (representing wildness) symbolises the possibility of a peaceful coexistence between nature and urban culture as nature "recaptures" the remnants of old industry. This coexistence also seems to prove that this new industrial wildness is no longer the opposite of urban culture but a part of it.

However, the design of this new type of industrial nature as a mixture of new wilderness and industrial remnants also incorporates the traditional concept of culture which designs images that contrast wild nature and culture by constructing "hybrids" of them.

Industrial nature is part of a long cultural tradition of these hybrids of nature and culture. There are many examples of cultural wildness as artificial wilderness in garden history that combine apparently wild vegetation and artificial objects representing culture or arts. This image of wildness as a mixture of an artificial wild nature and artefacts was developed in the Renaissance garden (Puppi 1993). An even more elaborate example is the mannerism garden of Sacro Bosco (Bomarzo, Italy) which combines (artificial) wildness, sculptures, buildings and even a zoo (Bredekamp and Janzer 1985).

Fig. 2. Wilderness as an hybrid of nature and art. Sacro Bosco, Bomarzo, Italy

The examples of wildness in garden history show that the cultural concept of wilderness includes wildness in the design. This design basically starts with the contrast between wilderness and culture. The contrast has been made readable by different aesthetic "designs" of wilderness throughout garden history. Wild vegetation became "designed wilderness" as well, for example, by integrating man-made elements that accentuate its natural character.

This design of wilderness represents the dialectical relationship of nature and culture that comes from the perspective of the prevailing culture. The model of industrial nature therefore reflects the current relationship of society to wild nature: "wildness" represents an autonomous development of nature in contrast to a totally controlled man-made environment.

Wilderness and the aesthetic idea of landscape

The basis for understanding the interpretation of wildness as an element of the peri-urban landscape is the perception of wilderness in the context of cultivated land. This interpretation is determined by cultural concepts of wilderness and landscape and their relationship.

However, approval of wildness as a part of industrial nature does not mean that elements of wildness are accepted in the peri-urban landscape. The acceptance of wildness is related to the perception and interpretation of landscape with its visual background. The traditional concept of landscape is dominated by the idea of cultivated land which represents more than just land use, it symbolises as well the unity of humans and nature (Seel 1991).

This concept of landscape can also integrate remnants of wild nature. However there can be some rejection of this wildness if it develops out of former farmland. This is because the first stages of natural vegetation development look like fallow land and therefore seems to represent the decay of cultivated land.

The change of landscape and its aesthetic concept

Growing dimensions of modern agricultural production distinguish the peri-urban landscape from traditional farmland. New urban elements interfere with the traditional image of a rural landscape. For this reason, the peri-urban landscape does not relate to the concept of cultivated landscape anymore: it is new scenery containing a mixture of both natural and artificial characteristics.

However this structural landscape has the potential for a different aesthetic perception. Wildness could be part of the new aesthetic concept because it represents a contemporary idea of spontaneous nature. Wildness as a symbol for autonomous nature could therefore produce aesthetic and symbolic contrasts to the functional urban elements in the landscape. This aesthetic concept would transform peri-urban landscapes into contemporary hybrids of culture and new wilderness.

Fig. 3. Artificial wilderness in an advertising brochure of a holiday park

Not only industrial nature, but also contemporary ideas of landscape work with images of wildness to create artificial nature. Historical ideals of landscape, like Arcadia as tamed wilderness, also worked with wildness as an aesthetic myth.

These examples show that there are aesthetic concepts necessary for integrating wildness into the landscape.

Design for new wilderness in the peri-urban landscape

The myth of landscape as cultivated land has also changed. The idea of the wild has become part of the cultural concept of landscape. This new wilderness will become a part of the peri-urban landscape as it meets not only the functional and ecological but also the symbolic needs of urban society. The perception of this symbolic meaning is the basis for the acceptance of wilderness in the peri-urban landscape.

To provide this perception, new wilderness with its wild vegetation must be managed in the design. Projects with wild vegetation have shown that acceptance of this kind of vegetation in urban public spaces is difficult to achieve just by explaining its ecological value. Its value is linked to an aesthetic perception which is based on a cultural concept of wilderness and not on the real object. The importance of design goals for providing access to wilderness has been demonstrated by the success of projects that include designed wild vegetation, e.g. the "Landschaftspark Duisburg Nord" (Keil 2002).

For this reason new wilderness must be designed so that it will meet the expectations of an aesthetic perception of the wild (even if this design is not visible). This design also has functional purposes but the aesthetic concept is needed so that the symbolic meanings of wildness become visible.

Strategies for the design of new wilderness in the peri-urban landscape

Strategies for the design of new wilderness in the peri-urban landscape must address the specific conditions there. For this reason, design methods with a minimum of expenditure and costs are most useful.

Some of these methods can be found in the design of models of wildness in garden history, landscape architecture and the arts. The design strategies of these models can not only help to develop design methods but ensure that wilderness elements will be accepted as part of a cultivated landscape.

Wilderness design as a contrast to regular pattern

A basic way to design new wilderness is to work with the visual perception of complex systems (see above). Therefore wilderness can be designed by establishing a visual contrast to a regular pattern or to an object with a regular form. This regular object or pattern, together with the wild vegetation, forms an aesthetic object. Another similar design method works by using the symbolic contrast of wilderness to cultural objects.

Contrast as a design principle for informal vegetation has a long tradition in garden history. Early examples can be found in the informal vegetation of the *bosco* of the Italian Renaissance gardens contrasting with their sculptures or regular forms (Comito 1993).

This principle is also used for the design of modern industrial brownfields, e.g. the design strategy for the Skulpturenwald Rheinelbe, in Gelsenkirchen, which uses modern sculptures by the artist H. Prigann as a contrast to the industrial woodland (Keil 2002).

Fig. 4. (left) Design by contrasting wild vegetation to a man-made object, Skulpturenwald Rheinelbe, Gelsenkirchen

Fig. 5. (right) Design by contrasting wild vegetation to linear edges, Natur-Park Schöneberger Südgelände, Berlin

Wilderness design through contrast with regular edges

Cultural models for designing new wilderness also demonstrate another principle of design by contrast, that between informal vegetation and a border. An aesthetic perception of informal vegetation is only possible if its border is linear or regular. Examples of this design concept are the linear edges of the informal vegetation of the bosquet areas of Baroque parks. However, it can also be seen in modern designs of urban brownfields. For example, the linear pathways of the Natur-Park Schöneberger Südgelände in Berlin form linear borders to the adjacent areas of wild vegetation.

A design for wilderness that establishes a contrast between the informal vegetation of new wilderness and man-made objects or a regular pattern also reflects the symbolic meanings of the cultural concept of wilderness. As well the design refers to the images that were developed in the arts and literature. It enables associations to history and therefore leads to new images of wilderness. Below, the question of whether design concepts are necessary for public acceptance of wildness in the (peri-urban) landscape will be discussed. The following section shows potential functions of new wilderness there.

Potential functions of new wilderness in the peri-urban landscape

The purpose of new wilderness in the peri-urban landscape is mainly determined by contemporary functions of the peri-urban landscape, which have little to do with the functions of the former rural landscape. They are a result of the function open space has for sustainable regional development in the future. This section shows some of these functions which have potential for new wilderness: places for ecological functions and leisure activities, local production of resources and the creation of an identity for the peri-urban landscape.

Ecological functions of secondary wilderness

The benefits of new wilderness for flora, fauna and habitats are well known, especially for urban brownfields. However, new wilderness could also be an effective strategy for managing the spontaneous vegetation of larger, floodplain areas of river systems.

Besides these functions for habitats, new wilderness could also serve for local sustainable production of resources, e.g. the production of timber as a substitute for oil products.

Wilderness to create an identity for the peri-urban landscape

The identity of the landscape is important for a positive image of an urban region. As the rural character of the peri-urban landscape is losing its dominance, it is important for regional planners to begin developing new identities for areas. Peri-urban park systems are one possibility. In most cases the strategy to develop this identity involves the conservation of historical buildings and methods of cultivation. However, a new identity could also be realised by a specific design concept for spontaneous vegetation.

Vegetation management could preserve part of the rural character and still allow for natural changes in vegetation. Even when fields turn fallow they will preserve some remnants of former cultivation. For example, a plantation of cherry trees will preserve its structure for a long time, no

matter what dynamic changes in the spontaneous vegetation occur. This strategy could help to develop a new identity from the former cultivation and character.
Concepts of land art and landscape architecture also show that minimal design intervention can make visible the specific history of a place or define its new character.

Wilderness for leisure activities

New wilderness in the peri-urban landscape can offer places for different leisure activities. These are different from commercial leisure activities as they offer the visitor a contemplative perspective on nature. From this perspective, the lack of any obvious design in the new wilderness equals a lack of information. On the other hand, this lack produces relaxation, which is a rare quality today because of the increasing information density in our environment (e.g. computer work and media).

However there are also active leisure activities which new wilderness can offer. It may provide space for "wild" activities that cannot be tolerated in urban parks (e.g. moto-cross or mountain biking). For this reason, new wilderness areas should not only be ecological sanctuaries but should be accessible to the public. At least some of them should be dedicated to "wilder activities". To curtail vandalism and littering, these areas need management including design goals, social controls and maintenance.

Wilderness elements as a pattern for urban development

One main issue in establishing sustainable urban regions is the management of land use. New wilderness areas could provide a large-scale pattern for the sustainable development of the whole of the peri-urban landscape. This pattern of new wilderness could help to manage urban development by "pre-structuring" its spatial area because woodlands provide strong obstacles to expansion while providing public open spaces. Spontaneous vegetation in new wilderness areas provides the basic structure for future parks in the (peri-)urban landscape.

A pattern of new wilderness for aesthetic perception of the peri-urban landscape

The main problem in the aesthetic perception of the peri-urban landscape is its complex structure. While the design of agricultural landscape is determined by the regular patterns of the borders of fields, pastures and hedges, the peri-urban landscape appears to be chaotic because it has no dominating structure and heterogeneous elements.

One strategy for developing an aesthetic perception of the peri-urban landscape could be to provide a readable structure. A regular pattern of new wilderness with linear borders would help as this would provide a reference to traditional cultivated landscape. This could also form a smooth visual background for the various elements of the peri-urban landscape and help to connect them visually.

Conclusion

The peri-urban landscape will be an essential element for leisure activities, sustainable urban development and the identity of urban regions in the future. This landscape will not be dominated by agricultural land but will have agriculture with different intensities. The peri-urban landscape will also have many new different urban uses and a constantly changing land use.

The process of urbanisation and changes in land use will lead to areas that temporarily lose their economic function. Concepts and effective methods must be developed to reuse and bring new qualities to these sites.

Management of natural vegetation, including wild vegetation, will be one approach for producing different kinds of open spaces in the peri-urban landscape. Sites with existing vegetation can be changed with little expenditure into places that provide ecological value and space for leisure activities. These new wilderness areas could also serve for regional recycling of disposable items.

On the scale of regional planning, new wilderness could also support sustainable urban development and create a new identity for the peri-urban landscape through design.

A structural pattern of new wilderness elements could provide a reserve of land for the management of natural resources. This pattern could also compensate for the chaotic appearance of a sprawling landscape.

The future will show whether the concept of new wilderness can describe all functions and forms of managed wild vegetation in the peri-urban landscape. However, new wilderness as a cultural concept shows clearly that an aesthetic and cultural perception of the natural development of vegetation is important. It is necessary to understand the mechanisms of acceptance and to develop design strategies for areas with wild vegetation.

This is the basis on which these areas can become important elements in the peri-urban landscape in the future.

References

Bredekamp H, Janzer W (1985) Bomarzo: Vicino Orsini und der heilige Wald von Bomarzo – ein Fürst als Künstler und Anarchist. Band I. Werner`sche Verlagsgesellschaft, Worms

Comito T (1993) Der humanistische Garten. In: Mosser M, Teyssot G (eds) Die Gartenkunst des Abendlandes. Deutsche Verlagsanstalt, Stuttgart, pp 33–42

Dettmar J, Ganser K (1999) IndustrieNatur Ökologie und Gartenkunst im Emscher Park. Ulmer, Stuttgart

Grossklaus G (1993) Natur-Raum: von der Utopie zur Simulation. Iudicium-Verlag, München

Keil A (2002) Industriebrachen: innerstädtische Freiräume für die Bevölkerung. Dortmunder Vertrieb für Bau und Planungsliteratur, Duisburger Geographische Arbeiten, Band 24, Dortmund

Körner S (2005) Nature Conservation, Forestry, Landscape Architecture and Historic Preservation: Perspectives for a Conceptual Alliance. In: Kowarik I, Körner S (eds) Urban Wild Woodlands. Springer, Berlin Heidelberg, pp 193–220

Losch S (1999) Nutzungsanspruch Wohnen, Nutzungsanspruch Landwirtschaft. In: Akademie für Raum- und Siedlungs-entwicklung/Akademie für Raumforschung und Landesplanung (eds) Flächenhaushaltspolitik Feststellungen und Empfehlungen für eine zukunftsfähige Raum- und Siedlungsentwicklung. ARL, Hannover, pp24–30; 104–110

Puppi L (1993) Kunst und Natur: der italienische Garten des 16. Jahrhunderts. In: Mosser M, Teyssot G (eds) Die Gartenkunst des Abendlandes. Deutsche Verlagsanstalt, Stuttgart, pp 43–54

Seel M (1991) Eine Ästhetik der Natur. Suhrkamp, Frankfurt am Main

Wulf A (1995) Neue Wege im Naturschutz. LÖBF Mitteilungen 4/95:35–42

Forests for Shrinking Cities? The Project "Industrial Forests of the Ruhr"

Jörg Dettmar

Design and Landscape Architecture, TU Darmstadt

Introduction

The central theme in urban planning in Germany currently is how to address the consequences of population decline. These consequences are already being felt in many places, in other areas they are fearfully expected. Shrinking cities and perforated cities are terms to describe the break up of urban structure in which new open spaces arise through the demolition of residential buildings and infrastructure facilities. Urban planning works to provide direction for an orderly retreat. It has long been unclear whether this has been successful (e.g. Arbeitskreis Stadterneuerung 2002).

None of this is completely new. It was already clear in the 1980s that these demographic developments would arise. Also, the old industrial regions of Europe – the Ruhr, for example – have several decades of experience with the shrinking process (Wachten 1996).

This marks a serious turning point for architects and urban planners, politicians and citizens who are accustomed to combining progress with growth and now must consider demolition and deconstruction. These aspects of decline have triggered helplessness and depression in recent years. Landscape architecture has been brought more sharply into focus in the search for new approaches to solutions. For one, this is for purely pragmatic reasons. When a building is torn down and a new built use has been ruled out over the long term, a new "open space" is created to be designed, to be developed, to be dealt with. Also, within landscape architecture, work with vegetation is the central element. Inherent in this is the integration of growth and decline. A dynamic and flexible work and design philosophy is far more necessary in landscape architecture than in architecture. This gives rise to the hopeful expectation that landscape architecture can also help in the development of process-oriented solutions which fea-

Kowarik I, Körner S (eds) Wild Urban Woodlands.

ture the necessary integration of growth and decline of urban structure (Dettmar and Weilacher 2003).

The rediscovery of open space, or of landscape architecture as the case may be, is thus explained. Communities, developers, and politicians are demanding, above all, new open spaces in the city – cost-effective, ecological, urban and attractive open spaces. This cannot be achieved with the traditional approaches and traditional building blocks of publicly financed green space. New strategies for the development of urban open space are being sought and tested in many places (see Rössler 2003). In this chapter, the experiment of the "Industriewald Ruhrgebiet" (Industrial Forests of the Ruhr) will be used to illustrate the solutions that were found in the largest former industrial, urban agglomeration in Europe.

The need for an experiment

In the mid-1980s, due to the increasing decline in mining and heavy industry, the Ruhr had reached a point at which ecological and cultural problems were becoming known, in addition to the area's substantial economic and social problems. The many conventional economic development policies that had been attempted had not shown any of the hoped-for success. The Ruhr featured the highest unemployment rates in North Rhine- Westphalia and in the former West Germany at over 15%. It was becoming increasingly clear that the negative image of the region, burdened by its industrial past, was becoming a decisive disadvantage. The image was that of faceless cities, insufficient visual landscape qualities, the legacies of industry in the form of ruins, abandoned areas, etc. During the years 1989 to 1999, the state government attempted, with the International Building Exhibition (IBA) Emscher Park, to provide a catalyst for a fundamental renewal of the areas of the Ruhr that had most been affected by industrialization—the central zone along the Emscher river. At the fore were those aspects that related to the creation of an attractive post-industrial cultural landscape. One aspect was the development of the Emscher Landscape Park, a new regional park with regional greenways, whose main features had already been considered in the 1920s (Schwarze-Rodrian 1999).

With the actual condition of the urban-industrial landscape as the starting point, remaining open spaces were to be connected as much as possible and some parts were to be further designed. The numerous abandoned industrial areas played a decisive strategic role in this. Due to the closure of coal mines, steel works, and numerous infrastructure facilities, e.g. factory railways, more than 1,000 ha of land lay abandoned at the end of the

1980s. A new built or otherwise profitable use for these areas appeared to be ruled out. The transformation into open space and the integration into the Emscher Landscape Park was a logical result that was made possible through the solid commitment of public funds within the framework of the IBA Emscher Park. In this way, a number of large new parklands of different types were developed during this period (Dettmar and Ganser 1999).

In principle, the creation of green spaces from former industrial land was not entirely new; at the end of the 1960s, for example, the Municipal Union of the Ruhr had already recultivated and revegetated a number of coal slag heaps on behalf of the communities and made them available as nearby recreational areas for the public. What was new was the consistent incorporation within the Emscher Landschaftspark (Emscher Landscape Park) and an intense engagement with the question of what design form was suitable for the post-industrial era. Numerous planning competitions and workshops were held. Significant projects included the Landschaftspark Duisburg Nord (Duisberg North Landscape Park), the Nordsternpark (North Star Park) in Gelsenkirchen or the Seepark (Lake Park) in Lünen (Schmid 1999).

It was also clear to the participants, however, that it would not be possible to actively redesign the greater part of the abandoned areas solely with the frugal financial conditions of the IBA phase. Furthermore, it was obvious that the future would bring large financial problems for almost all of the communities, above all the difficulties which come with the long-term maintenance of green space.

In addition, a number of those responsible within IBA Emscher Park, Ltd. were somewhat dissatisfied with the results of the new park planning. Despite all efforts, too few truly regionally specific designs had been developed. The system of planning competitions, landscape architectural design, implementation plans, and recreation concepts allowed too much of the industrial-historic, aesthetic and ecological substance of the abandoned areas to be lost. In the best cases, an attractive new park was envisioned that was no longer site-specific, but could have been built anywhere. Too many people in the system were responsible only for aspects of the whole, no one was there from the beginning to the end. At the end of the planning and construction process, the community open-space department would remain and have to find a solution for the upkeep with scarce financial resources. This alone would lead inevitably to a loss of quality.

The other alternative, to leave the spaces to themselves and do absolutely nothing, also offered no sensible prospect for a solution. On many sites that had been abandoned for longer periods of time, one could observe how the ecological and aesthetic qualities had clearly been affected, for example, by trash disposal or illegal motorbike use. In addition, a few owners used the sites for the temporary storage of soil und other substances or rented them out for a small fee to construction companies, moving firms, etc. Furthermore, entrance to the sites remained illegal and was only possible and of interest for a part of the public. The fundamental reasons for these issues were the absence of social controls and the potential safety risks of the ruins, mine shafts and brownfields.

The basic approach of the project

Under conditions such as these, a new approach was needed. Additionally, it was already clear that demographic changes would sooner or later lead to the availability of a substantial number of previously built areas throughout Germany. For this reason as well, model strategies needed to be developed. The most important goals associated with a new approach can be classified as follows:

- Consider the increasingly difficult financial situation of the community.
  - Cost-effectively transform areas in the city that no longer have an economic use into usable green spaces of different types that require very little upkeep.

- Consider alternative methods to avoid the blandness and loss of quality that is involved in the "normal" planning process when transforming an abandoned area into a park.
  - Offer solutions, usually absent in normal conversion approaches, that consider the specific aesthetics, the species and biotope conservation potential and the dynamic development potential of nature in abandoned areas.
  - Protect the authenticity of the abandoned areas during redesign; preserve the natural potential that is always an expression of the industrial history of a site.
  - Understand industrial nature as a counterpart of industrial culture.

- Make the areas available to the public.
  - Find intelligent solutions to the problem of liability which forces the owners of properties, especially communities, to eliminate even the slightest risk and thereby brings about a homogenization.
  - Set up social controls.
  - Make contact with nature possible in the city by creating or conserving natural areas that are appropriately accessible.
- Augment forests in forest-poor urban areas such as the Ruhr.

The approach, therefore, was not to elaborately redesign the abandoned areas based on conventional planning, but rather, in principle, to allow the natural development, succession to have free rein. The starting point was the realization that all stages of succession up to and including the mature forest feature ecologically interesting and aesthetically appealing elements (Dettmar 1999, 2004).

What was being sought was something best described as "nurtured development". The responsibility for a site would no longer be divided between planning and realization phases, but rather would be anchored in one person. This person would be present as much as possible at the site and thereby secure a certain social control.

The implementation

The idea of nurtured development, of maintenance and cultivation, quickly brought the historic image of the forester into the minds of the parties involved in the IBA Emscher Park. The employment of foresters in the abandoned areas is a logical step for the process of succession towards the forest, a process which happens relatively quickly on most sites. Federal and state forest regulations allow for even early stages of growth without a dominant woody layer to be defined as forest development areas. The formal transformation of a former industrial site into a forest at the level of land use planning will secure this status. The forest classification has significant advantages for the owners, at least when the change in the property valuation does not present economic problems. As long as the site is identified as a normal production forest and not explicitly as a recreation forest, there are lower standards for liability than would be the case for a park. This is, at least, the estimation of lawyers of the Forestry Administration in North Rhine-Westphalia (NRW).

In 1996, the experiment was begun under the title "Restflächeprojekt" (Remnant Land Project) in collaboration with the State Forestry Administration NWR. Work was initially begun on three core sites in the central Ruhr on old abandoned industrial sites, areas for which, for the most part, no new built use was expected. The Grundstückfonds NRW (a public fund of NRW for purchasing property) acquired the sites for the State of North Rhine-Westphalia with the objective of developing new commercial space in certain areas; the remainder of the land was to be transformed into green space.

The remainder included parts of the former Rheinelbe coal mine in Gelsenkirchen (ca. 40 ha that had been decommissioned in the 1920s; parts were to be developed as the Rheinelbe Industrial Park/Science Park), parts of the former Alma coal mine in Gelsenkirchen (ca. 30 ha, decommissioned in the 1960s, with significant areas of brownfields), and parts of the former Zollverein Shaft XII coal mine and Zollverein coking plant in Essen (ca. 40 ha, decommissioned in the 1980s, of significant industrial historic preservation value). An extensive description of the sites is available in Rebele and Dettmar (1996).

The core sites were integrated into the project on the basis of a *Beförsterungsvertrag* (forest maintenance contract) with the Forestry Office of Recklinghausen according to the state forestry regulations of North Rhine-Westphalia. The *Beförsterungsvertrag* provides for private forests to be overseen by state forestry offices. In this way, taxes of only a few euros per hectare on the land are due.

In total, three employees of the State Forestry Office NRW were made available for the project. On the site of the Rheinelbe in Gelsenkirchen, a forestry station was created through renovation of an old switch house of the coal mine.

During the test phase from 1996 to 1999, a total of approximately 500,000 Euros were available for the project as start-up money from the EU in combination with funds from the Emscher Lippe Ecology Program of the State of North Rhine-Westphalia; about 70% were used for the renovation of the forestry station. The remaining money was available for provisions for the site. The labor costs of the foresters are paid by the forestry administration.

At the same time, a expert advisory board was assembled, with representatives of almost all of the institutions that are relevant to the project (ministries, communities, property owners) as well as scientists. The board determines the basic concepts for the development of the project and the integrated areas. Decisions regarding the use of funds fall to this group as well. Furthermore, the board is intended to assist the foresters with problems that arise with the institutions or with the forestry administration.

For the core areas, development concepts were worked out by the author that were then adopted by the advisory board. These served as guidelines for the maintenance of the areas for the foresters during the first years of the project. In this way, the employees of the forest administration could be informed and made aware through their work with the succession forests and with the abandoned areas and their particular ecology.

In addition, ideas were developed about necessary measures for development and providing access to the sites (path design), safeguarding against dangers (brownfields, accident-prone locations), species and biotope conservation (promotion of individual species), and special public uses (keeping some sites open). Just as important was defining the areas in which no interventions would be made.

After a period of learning, the foresters continued to develop their own ideas for managing the sites, coming closer to the ideal of nurtured development. A prerequisite for this, however, is a stable workforce.

The offer of regular guided tours through the "industrial nature" became a very important part of the work of the foresters. The demand has steadily increased. To this end, the forest station has been outfitted with educational, seminar, and class rooms for school children, preschoolers and other groups.

Table 1. Overview of the number of participants on guided tours through the Industrial Forest Project from 1998–2003 (Source: Statement of the Rheinelbe Forest Station in June 2004)

Year	Number
1998	1,800
1999	3,490
2000	2,800
2001	3,100
2002	3,300
2003	3,500

Within the framework of the IBA Emscher Park, sculptural works of artists were integrated into the two core areas of the Rheinelbe in Gelsenkirchen (nine sculptural works of Herman Prigann) and the Zollverein in Essen (five sculptures of Ulrich Rückriem) (see Dettmar 1999; Prigann 2004). In particular, Herman Prigann with his work in Rheinelbe, attempted an artistic interpretation of the transformation of the abandoned areas through natural succession and of the appropriation of the sites by humans. At the same time, the artistic installations help to change the visitor's perception of the site (for a detailed description, see Strelow 2004). Those involved in the project anticipated that the art would bring value to the abandoned areas and therefore lead to a greater public acceptance.

Certain areas were intentionally incorporated into the project for limited time periods. This was true, for example, for part of the site of the Rheinelbe coal mine in Gelsenkirchen. There, as planned from the beginning, a part of the Rheinelbe Industrial Park/Science Park was to be developed. This was intended to show that intermittent management of the abandoned areas was also possible through the project. This is of interest for owners who still anticipate built development on their property in the medium to long term.

It was expected from the start that private land would be integrated into the project. The largest part of the abandoned industrial areas in the Ruhr remain in private hands, especially large firms like Thyssen-Krupp or Deutsche Steinkohle. It would not be possible to acquire all of these sites through public funding. When no other interested parties are to be found, the project can be an interesting partner for firms. In the meantime, a number of appropriate sites have been incorporated; in addition to the *Beförstungsvertrag*, conventional lease agreements have also been used.

Table 2. Overview of the sites of the Industrial Forest Project of the Ruhr. (Source: Statement of the Industrial Forest of the Ruhr, Recklinghausen Forest Office, June 2004)

Site	Area (ha)	Location
Rheinelbe coal mine	42	Gelsenkirchen
Emscher-Lippe 3/4 coal mine	34	Datteln
Alma coal mine	26	Gelsenkirchen
Waltrop coal mine	26	Waltrop
Zollverein coking plant	21	Essen
Graf Bismarck coal mine	20	Gelsenkirchen
Zollverein Shafts I, II, VIII, XII coal mine	20	Essen
Hansa coking plant	20	Dortmund
Chemische Schalke chemical factory	13	Gelsenkirchen
Victor 3/4 coal mine	12	Castrop Rauxel
Constantin 10 coal mine	8	Bochum
König Ludwig 1/2 coal mine	2	Recklinghausen
Total Area	244	

Further areas totaling a few hundred hectares are currently in the process of being incorporated.

The "Restflächenprojekt" successfully closed out its five-year test phase in 2000 and was established as a permanent project of the State Forestry Administration NRW in 2001. Thereafter the project was operated and further developed by the Forestry Administration through the Recklinghausen Forest Office and through the Rheinelbe Forest Station in Gelsenkirchen.

This has all worked out well when measured by the number of regular visitors and the number of tours. Through public works, the Forestry Administration was able to build an important foundation directly in the cities of Gelsenkirchen and Essen. The attractiveness of the sites to children and youth is especially important; they find a much greater degree of freedom there than in most urban open spaces. On these sites, truly direct contact with nature takes place (see also Keil 2000).

Applicability

What can be applied in other cities or regions? Clearly such a fortunate case as the IBA Emscher Park with its start-up financing to serve as a catalyst occurs only rarely. Of course, the foresters entail labor costs—nevertheless, the development and maintenance of these sites is many times more cost-effective than conventional green spaces (see Dettmar 1997).

It is conceivable that a corresponding plan for care/custody could also be constructed from labor market projects, citizens' or nature-conservation organizations, residents, or other kinds of volunteers. What is needed is a coordinating and supervisory site, but why shouldn't this be located in the forestry or open-space department? Enthusiastic employees are, however, a prerequisite for this idea. A certain level of knowledge about succession is also necessary, in order to learn that one can withstand the increasing wilderness, that the urgent need to intervene can be held in check. The essential features of structural development and the necessary safeguarding against danger must also be carefully determined. The issues of liability certainly can not be fundamentally neglected, but in the forest there is more leeway.

There are probably more lessons to be found in this approach. Perhaps the most valuable contributions to the future of landscape architecture are to be found in the ideas, suggestions, and solutions for the function and design of residual open spaces. How does one develop a sustainable and attractive urban landscape from the abandoned landscapes of endless suburbanized developments? Can one succeed in creating a sustainable and attractive urban structure within the perforated urban structure using the abandoned lands that arise when cities shrink?

Clearly the demands of the suburban growth zones and of shrinking cities are very different at first glance. In one case, there is great economic pressure and need for space, in the other there is shrinking and retreat. From a structural perspective, however, under both conditions an urban space consisting of a patchwork of built and open spaces arises.

When urbanization processes are viewed fundamentally, growth and shrinking belong together. The further development of urban industrial society has produced a kind of "total industrial landscape", as Sieferle (1997) has described it. This landscape is characterized by a constantly increasing flow of information and a universal disposability of materials and is based on an increasing use of energy. Consequently, an apparently individual and fleeting pattern of differentiation appears, a unity of variety and monotony based on the same universally available fashions, building styles, architecture, and garden designs with the corresponding merchandise, building materials, and garden center products. In contrast to the old cultural landscapes, no new permanent, truly recognizable style emerges. This would require far more development time and regional isolation. The one characteristic that remains is constant change. This mobilized stylelessness is the one overarching feature of our urbanized landscape, the only constant is the permanence of change (Sieferle 1997). This applies as a functional principle to the entire space, independent of suburban growth or urban shrinking. It also generally applies independently of the degree of development, and independently of the historic categories of urban and rural.

We attempt to guide processes and to achieve a design through planning that will eventually create an economically functional, attractive, liveable, functioning urban or landscape structure. With regard to the intercity structures of suburban spaces, most experts believe that, thus far, we have not succeeded. The hope of architects and urban planners is centered on the potential to structure, to organize, and to give new identity to the intercity areas through open space (Bächthold 1995; Sieverts 1997). Here as well, open space experiences an urban "flight of fancy" (Lohrberg 2002).

The different approaches to regional parks in Germany (the Emscher Landscape Park in the Ruhr, the Rhein-Main Regional Park, the Stuttgart Green Neighborhood, the Filder Raum Regional Park, etc.) operate according to this strategy. With this, planning for regional greenways and systems for open space connections are provided for. In doing so, planning follows the most common goals of safety, care, order and design.

If one follows the analysis of Sieferle, the attempt to create order out of chaos is understandable, but doomed to failure from the start. Permanent change ultimately excludes stable patterns of order. The recourse to typical landscape elements of the pre-industrial era (the Rhein-Main Regional Park) or the aesthetic staging of the likewise bygone industrial landscape (the Emscher Landscape Park) are only integrated elements in a mobilized landscape, in which these museum-like or symbolic islands only emphasize the totally constructed character of the landscape.

Principles of organization ultimately originate from fears or from the need for harmony; returning to what is known is understandable. What happens when this fails? Because we still have no clear vision of the structure, function, design and qualities of the mobilized, urbanized landscape of the Information Age, much energy is currently being expended to study and to understand the existing conditions and the mechanisms by which the existing conditions arose (Lootsma 2002). We must examine to what extent our perceptions, shaped as they are by historic images and representations, allow us to perceive potential qualities or organizational patterns in new structures (Dettmar and Weilacher 2003).

During shrinking processes as well, one attempts, through planning, to retreat in an orderly fashion, to avoid allowing merely accidental factors to determine the makeup of the new urban structure. Where and at what scale demolition will occur, where new green spaces will be arise and how these can be sensibly joined together is a process that must be guided (Giseke 2002).

Whether it is the shrinking of the urban structure from the era of industrial expansion (e.g., in Leipzig's Osten) or large developments at the edges of cities (e.g., in Berlin's Marzahn) is unimportant—the newly created open spaces must give a new organizational pattern to the whole. Forest edges and walls of trees form the new edges of spaces when buildings are broken apart (Giseke 2002). Behind these changes is fear, fear of the disintegration of a beloved urban structure, of the end of the traditional European city.

Wilderness in abandoned areas will only be accepted as long as it develops within a specified framework, as long as it fits in with the planned pattern of new open spaces. "In many places, open space in disintegrating cities is given the task, not of bringing wilderness to the city, but rather of properly maintaining the continuum of urban development and public social space. In other words: urban planning through landscape" (Becker and Giseke 2004).

What role will abandoned areas, increasing wilderness, and succession forests have in the future?

Abandoned areas can provide a building block for urban open spaces. They certainly aren't without cost, but as the Industrial Forest Project shows, they are much more cost-effective than other public green spaces. The option of potential re-use or new construction, should the social and economic conditions change, makes abandoned areas attractive as well. Under certain circumstances they offer great ecological, aesthetic and social qualities. As islands of transformation determined by nature, they can offer a different kind of permanent change within the constant transformation of the total urban landscape, while always presenting site-specific character. This is more true the more time they have to develop. Particularly for the development of children and youth, a touchable, usable, uncontrolled, wild experience with nature is important (Gebhard 1998).

Abandoned areas can also serve as land reserve for the creation of energy, water and nutrient recycling systems in urban landscapes. One must only bring to mind the example of semi-natural rainwater management in cities (Londong and Nothnagel 1999). From a sustainability perspective, creating decentralized wastewater systems with plant purification systems, biomass production, and biogas usage is sensible within urban landscapes on suitable open spaces (see Ripl and Hildmann 1997). A co-existence of traditional green spaces, succession forests, drainage infiltration areas and plant water-treatment systems is conceivable. These are likely important building blocks of the urbanized landscape in the Information Age.

Cities were historically seen as places free from the dangers and risks of nature. The cultural break between the city and the landscape began with our estrangement from direct food production and from the discovery of the landscape by "emancipated" humans. Landscape became a synonym for nature and increasingly took on aesthetic symbolism. When "nature" found an entry into the city it was in the civilized aesthetic, staged form of gardens and parks, though with very different visions of nature during different cultural eras. In each case, a "wild" spontaneous "nature" suited to the urban conditions of the city was not a symbolically ideal nature, but rather a profane expression of urban reality. As well, such nature was an expression of a city not functioning perfectly.

At least in Germany, wild abandoned areas that exceed a certain scale engender strong psychological fear. This is true for cities, but also generally for cultural landscapes shaped by agriculture. In cities, this clearly arises from a cultural historical basis, from the consequences of wars from the Thirty Years' War with its deserted towns to the destruction of World War II.

This must all be understood if one is to see abandoned areas as integrated building blocks of urbanized landscapes and not as an unavoidable evil. With the end of the Industrial Age, we have arrived at a point at which we must question and examine the function, the perception and the design of open space in the completely urbanized society and landscape. A central issue in this is the question of the human understanding of nature and our relationship with nature in the Information Age.

A core element of a new relationship with nature seems to be a stronger focus on the development principles of nature and less on a particular stage of development. In this sense, abandoned areas can be places of learning and experiencing a transformation guided by nature within a mobilized landscape.

References

Arbeitskreis Stadterneuerung an deutschsprachigen Hochschulen, Institut für Stadt- und Regionalplanung der TU Berlin (eds) (2002) Jahrbuch Stadterneuerung. TU Berlin. 355 pp

Bächthold HG (1995) Landschaft – die neu entdeckte Dimension der Raumplanung ? DIS, Heft 123

Becker CW, Gisecke U (2004) Wildnis als Baustein künftiger Stadtentwicklung? Garten + Landschaft 2/2004:22–23

Dettmar J (1997) Gestaltung der Industrielandschaft. Garten + Landschaft 6/97:9–15

Dettmar J (1999) Wildnis statt Park? Topos 26:31–42

Dettmar J (2004) Ökologische und ästhetische Aspekte der Sukzession auf Industriebrachen. In: Strelow H (ed) Ökologische Ästhetik. Theorie und Praxis künstlerischer Umweltgestaltung. Birkhäuser, Basel, pp 128–131

Dettmar J, Ganser K (eds) (1999) IndustrieNatur – Ökologie und Gartenkunst im Emscherpark. Ulmer, Stuttgart

Dettmar J, Weilacher U (2003) Baukultur: Landschaft als Prozess. Topos 44:76–81

Gebhard U (1998) Stadtnatur und psychische Entwicklung. In: Sukopp H, Wittig R (eds) Stadtökologie. Ein Fachbuch für Studium und Praxis. Gustav Fischer, Stuttgart, pp 105–121

Giseke U (2002) Urbane Freiräume in der schrumpfenden Stadt. Der Architekt 8:44–46

Keil A (2000) Industriebrachen – innerstädtische Freiräume für die Bevölkerung. Mikrogeographische Studien zur Ermittlung der Nutzung und Wahrnehmung der neuen Industrienatur in der Emscherregion. Ph.D. thesis, Gerhard-Mercator-Universität Duisburg, 347 pp + Appendices

Lohrberg F (2002) Landschaftsarchitektur als Städtebau. Garten + Landschaft 10:10–12

Londong D, Nothnagel A (1999) Bauen mit dem Regenwasser. Aus der Praxis von Projekten. R. Oldenbourg Industrieverlag, München, 236 pp

Lootsma B (2002) Die neue Landschaft. Der Architekt 8:20–23

Prigann H (2004) Skulpturenwald Rheinelbe, 1997-2000, Industriebrache, Gelsenkirchen, Deutschland. In: Strelow H (ed) Ökologische Ästhetik. Theorie und Praxis künstlerischer Umweltgestaltung. Birkhäuser, Basel, pp 132–137

Rebele F, Dettmar J (1996) Industriebrachen - Ökologie und Management. Eugen Ulmer, Stuttgart, 188 pp

Ripl W, Hildmann C (1997) Ökosysteme als thermodynamische Notwendigkeit. In: Hübler K-H, Brahe K, Müller S, Krusewitz K, Bechmann A, Ripl W, von Wissel C, Hanisch J (eds) Beiträge zu einer aktuellen Theorie der räumlich-ökologischen Planung. Verlag für Wissenschaft und Forschung, Landsberg

Rössler S (2003) Schrumpfung – neue Chancen für die Freiraumentwicklung? Forderungen an den Stadtumbau aus der Sicht der Freiraumplanung. Stadt + Grün 11:14–18

Schmid AS (1999) Neue Gartenkunst. In: Dettmar J, Ganser K (eds) IndustrieNatur – Ökologie und Gartenkunst im Emscherpark. Ulmer, Stuttgart, pp 57–129
Schwarze-Rodrian M (1999) Der Emscher Landschaftspark – Ansätze und Strategien zur Gestaltung der Industrielandschaft. In: Dettmar J Ganser K (eds) IndustrieNatur – Ökologie und Gartenkunst im Emscherpark. Ulmer, Stuttgart, pp 38–54
Sieferle RP (1997) Rückblick auf die Natur. Eine Geschichte des Menschen und seiner Umwelt. Luchterhand, München, 233 pp
Sieverts T (1997) Zwischenstadt – Zwischen Ort und Welt, Raum und Zeit, Stadt und Land. Birkhäuser, Braunschweig
Strelow H (ed) (2004) Ökologische Ästhetik. Theorie und Praxis künstlerischer Umweltgestaltung. Birkhäuser, Basel, 255 pp
Wachten K (ed) (1996) Wandel ohne Wachstum? Stadt-Bau-Kultur im 21. Jahrhundert. Katalog zur VI. Architektur Biennale Venedig 1996. Vieweg, Braunschweig, 212 pp

Post-Industrial Nature in the Coal Mine of Göttelborn, Germany: The Integration of Ruderal Vegetation in the Conversion of a Brownfield

Justina Drexler

Department of Landscape Architecture and Spatial Planning, TU München

Context

This paper describes a design project that illustrates the integration of ruderal vegetation in the conversion of a mining brownfield in the Saarland, Germany. The coal mine of Göttelborn is located on the periphery of the state capital of the Saarland, 15 km north of Saarbrücken. Göttelborn is situated on the northern outskirts of the peri-urban Saarkohlenwald woodlands and represents one of the main industrial zones in the Saarland. Unlike other industrial regions, the Saarkohlenwald offers a sparse and rural settlement structure (Fig. 1).

Fig. 1. The landscape of Saarkohlenwald (photo: Wendl)

In the context of an industrial heritage network organised by the Commission "IndustrieKultur Saar" of the state government of Saarland, the pit frame of Göttelborn presents a highly visible landmark and stands as a testimony to the latest industrial architecture (IndustrieKultur Saar 2000).

Not only are the prominent buildings of Göttelborn a part of the industrial heritage, but also the huge changes in the landscape created by the mining activities. Some of these changes, such as the former sinking pond

Kowarik I, Körner S (eds) Wild Urban Woodlands.

and the mine slagheap are now outstanding landmarks. The sinking pond, today filled in with mine spoil, represents a black PLAIN in contrast to the surrounding landscape of the Saarkohlenwald woodlands. The mine slagheap, 50 m above the pond, is at the same height as the natural westward ridge and constitutes a landscape sculpture.

In conjunction with another sinking pond, the Kohlbachweiher, which is 120 m below and located to the south of the mine slagheap, these landscape formations have modified the original shape of the ancient valley (Deutsche Steinkohle AG 2000; Fig. 2). The changes in the landscape by mining activities result a spatial arrangement of "MOUNTAIN–PLAIN–VALLEY".

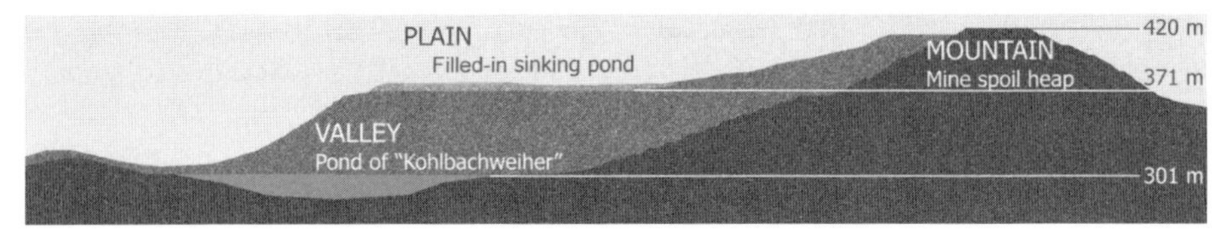

Fig. 2. Section of the landscape formation MOUNTAIN—PLAIN—VALLEY at Göttelborn

Concept

Post-industrial nature

Many mining areas in the Saarland have been recultivated with woody plants and are therefore hardly recognisable within the surrounding landscape of the Saarkohlenwald woodlands. The goal of this project is to preserve the landscape formations, such as the mine slagheap and the filled-in sinking pond as a testimony to the industrial history and to document the huge human impacts on the landscape. The new landscape offers a unique appeal and a characteristic morphological language based on the industrial history. The aesthetic does not correspond with the usual images of "ideal nature" (Dettmar 1999). In Göttelborn, a designed intervention will emphasise the potential of "post-industrial" nature, i.e. vegetation established on industrial sites, known as nature of the fourth kind as defined by Kowarik (1992, 2005). The intervention will present a special stage for discovering nature and experiencing the landscape through its characteristic aspects, e.g. the original barenness and changes brought about by vegetation (see more examples for this method of treating brownfield in Grosse-Bächle 2005; Henne 2005; Körner 2005).

The phenomenon of self-similarity

The black PLAIN of the former sinking pond resembles a lunar landscape and forms a contrast to the surrounding landscape of the Saarkohlenwald woodlands. The completely open space of about 20 hectares appears as a huge plain lacking human scale, structured only by the water-areas.

Fig. 3. Panoramic view of the former sinking pond

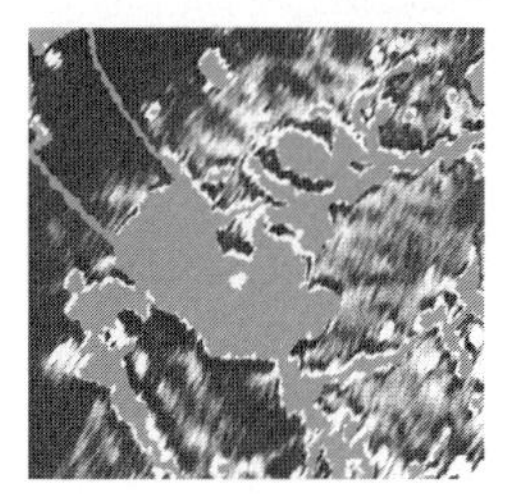

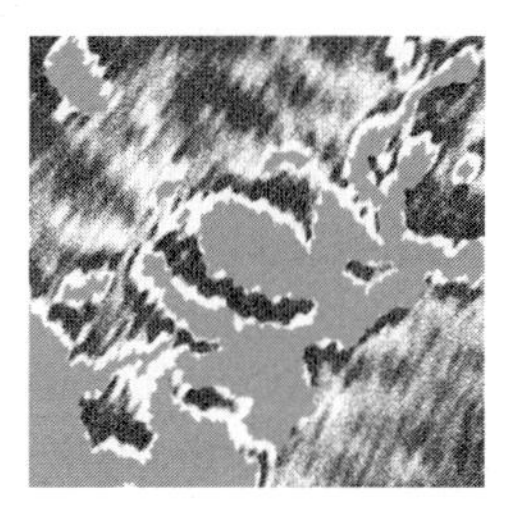

In the contours of the water areas of the PLAIN, the phenomenon of self-similarity or fractals, described by the mathematician B. Mandelbrot, can be found: basic structures recur regardless of the focus or the scale of the field of view (Mandelbrot 1987). The outlines of the water areas recur again and again in the shapes and contours.

The principle of self-similarity can be applied at larger scales, too. When observing the coloured aerial photograph of the PLAIN, it is not possible to distinguish whether the abstract shapes belong, for example, to the Lakeland area of Finland or to the PLAIN of the Göttelborn coal mine. Furthermore, the ruderal vegetation of the industrial landscape is similar to a forest if viewed at a larger scale.

Fig. 4. a, b Recurring shapes in the outlines of the water areas

The black PLAIN can be considered as a stage for the "actors" that exhibit the phenomenon of self-similarity in their outlines: coal-related material may be heaped and the development of ruderal vegetation may be accelerated by sowing.

Areas of ruderal vegetation show the same phenomenon of self-similarity in their outlines as the water areas. The bulk heaps that mark the boundary of the water areas are further existing elements within the PLAIN that produce the same formation.

Fig. 5. The boundary of the water formations of the PLAIN

Fig. 6. Contours of ruderal vegetation (photo: Hamann)

Visitors experience the phenomenon of self-similarity visually. The shapes of the different elements are made perceptible through optical aids comprised of lenses that enlarge the landscape: A catwalk over the PLAIN with fixed magnifying glasses facilitates an overview at a higher level. On the ground level of the PLAIN, lenses are installed as well.

Through this unusual perspective, the aesthetic effects of post-industrial nature are pointed out to the visitor. Post-industrial nature is contemplatively perceptible as landscape.

The phenomenon of change

A further aspect of the PLAIN of the sinking pond is the phenomenon of change. All the elements, the existing water areas supplied by rainwater, as well as the added elements like the bulk heaps and the vegetation, are characterised by change. The vegetation varies in expanse, coverage and shape. More changes become perceptible through vegetation management. In addition, the process of erosion changes the heaps of the coal material.

Levels of perception

The contemplation of the PLAIN is possible from different levels of perception: the MACRO-view offers an overview of the complete PLAIN whereas the MICRO-view allows the viewer to ZOOM into details.

MACRO-View

In the buildings of the coal-mine plant, the phenomenon of change may be documented, visualised and perceived by the visitor. The design in the buildings' interiors includes a multimedia presentation of the PLAIN with interactive screens, projections of aerial photographs, and additional photography. The PLAIN is partly reduced to an abstract graphic object worthy of contemplation.

The vantage points on the top of the coal-mine buildings, especially the look-out platform at the 90-m-high pit frame IV, and the highest plateau of the mine slagheap facilitate the immediate MACRO-view, the total overview of the complete PLAIN. Directly over the PLAIN, the 5-m-high catwalk allows an unmediated perception of the MACRO-view. Furthermore, alienating optical aids (lenses, colour filters, fish-eyes) are provided in "lens boxes".

MICRO-View

Analogously to the MACRO-view, the MICRO-view is established and alienated. In the lens boxes of the catwalk, Fresnel lenses also facilitate the ZOOM into the formations of the PLAIN and allow for discovery of the "form within the form", in accordance with the phenomenon of self-similarity.

On the ground level of the PLAIN, viewpoints with lenses are installed. Magnifying glasses improve the MICRO-view of the different elements such as vegetation, bulk heaps and water formations, and also act as EYE-CATCHERS focussing the view of the observer to a micro-level that is usually not perceptible.

Through the MACRO- and MICRO-views, the ruderal vegetation typical of industrial sites is visualised as an element of design, and an impression of the dynamics of nature is presented.

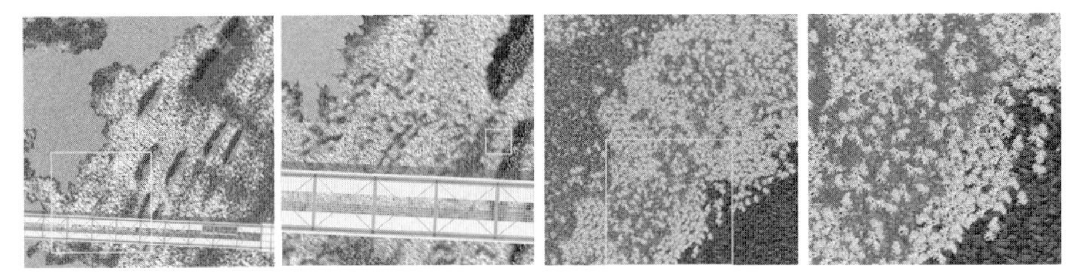

Fig. 7. Zoom into detail

Organisation: zonation

The water formations of the PLAIN provide the design framework. They structure and divide the area. Within this framework, the catwalk, steel-paths and lens viewpoints present fixed points.

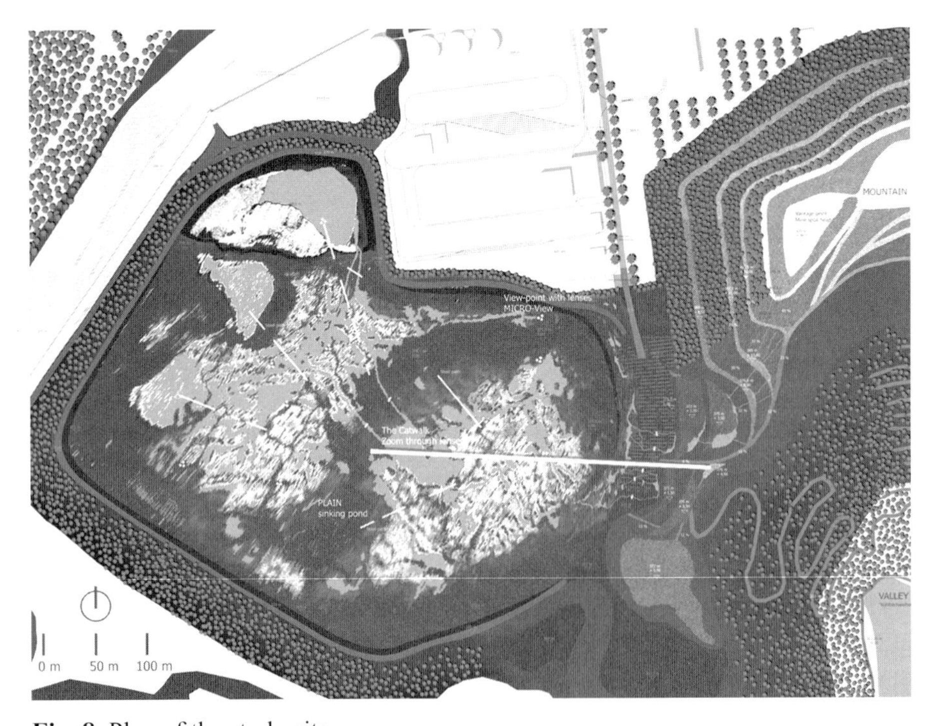

Fig. 8. Plan of the study site

The focus of the vegetation management is directed at primary stages of succession—from mosses to annuals to perennials—in order to preserve the impression of the PLAIN's vast expanse. The vegetation management creates a pattern of different vegetation types and divides the site according to the framework of the existing water areas.

The mine spoil areas which do not have any water formations are without vegetative cover. Next to the areas without vegetation, the natural vegetative development is artificially slowed.

The outlying area of the water formations is left to the natural succession of pioneer vegetation. In the mine spoil areas within the water formations, the vegetation development is accelerated through restoration treatments such as sowing and the application of an appropriate substrate. The different forms of vegetation are marked by several bulk heaps intended to provide additional structure to the area.

Deceleration of vegetation development

The vegetation development is slowed by heaping blast-furnace slag. This material offers an unfavourable growing medium with extremely alkaline conditions which have a toxic effect on plants, especially on deep-rooting species. These extreme habitat conditions lead to a low rate of succession. First mosses, e.g. *Ceratodon purpureus* and *Bryum argenteum*, colonise the slag substrate (Rebele and Dettmar 1996). Gradually the pH value is decreased through eluviation under the influence of rain. The substrate will then be populated by ruderal species like *Chenopodium botrys* and *Arenaria serpyllifolia* (Punz 1989).

Acceleration of vegetation development

For the acceleration of vegetation development, seeds of *Senecio* species (*S. viscosus, S. vernalis*), *Reseda luteola* and *Inula graveolens* are brought into the mine spoil ground of the former sinking pond.

This process improves the succession conditions and offers a natural source of seeds and other diaspores for further succession. The yellow-flowering colour of these species corresponds to the typical colour of pioneer vegetation on industrial sites (Dettmar 1999). When viewed from the near distance or by zooming, the various textures, flower shapes and habits of the plants are visible. Areas of accelerated natural vegetation development are marked by coal heaps.

The applied restoration treatment, investigated by M. Jochimsen, involves the application of small amounts of sand and fertilizer as required to increase the plant coverage (Jochimsen 1987). At first the seeds are sown in limited linear areas organised in a rational arrangement. Gradually the contours change into amorphous structures as the plant cover increases until the original sowing area cannot be defined anymore.

Natural succession of pioneer vegetation

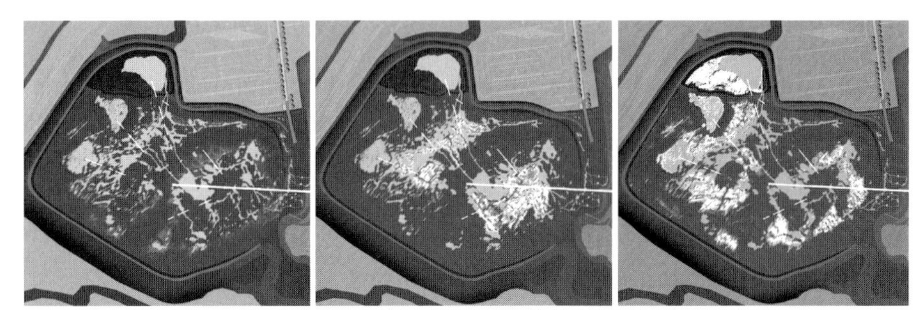

Fig. 9. Areas with deceleration (left) and acceleration (center) of vegetation development and natural succession of pioneer vegetation (right)

Finally the pioneer species are replaced by the communities of secondary succession. A herbaceous vegetation with a reduced number of species results, e.g. *Hypericum perforatum* and other perennials (Jochimsen 1991). Based on vegetation mapping of similar sites in the Saarland, the following species are also to be expected: *Conyza canadensis*, *Epilobium lanceolatum*, *Saxifraga tridactylites* and others (Maas 2002). Areas of natural pioneer vegetation are marked by coke heaps.

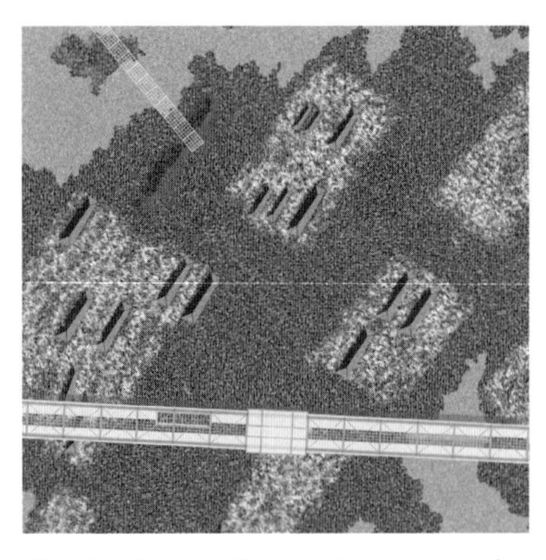

Fig. 10. Restoration treatments: sowing

Fig. 11. Expansion of plant cover

Changes in vegetation

Vegetation management is a method of design combining human impacts and natural processes and offers an impression of the dynamics of nature.

Impacts/disturbance

In order to avoid the colonisation of woody plants like birches, the plant cover is partially removed at periodic intervals. Once again, mine spoil is heaped on the site and pioneer habitats are created.

These areas will either be sown with seeds of pioneer species or will be left to the natural succession of the existing vegetation which provides plenty of seed sources. The changing system of vegetation management creates a pattern of different vegetation stages. Through the removal of the plant cover, the contour lines of the vegetation areas are once again clearly structured and linear, but will change over time into an amorphous structure. In addition, the original linear shape of the heaps dissolves gradually due to erosion and removal of the plant cover. The result will be a pattern of different stages of erosion.

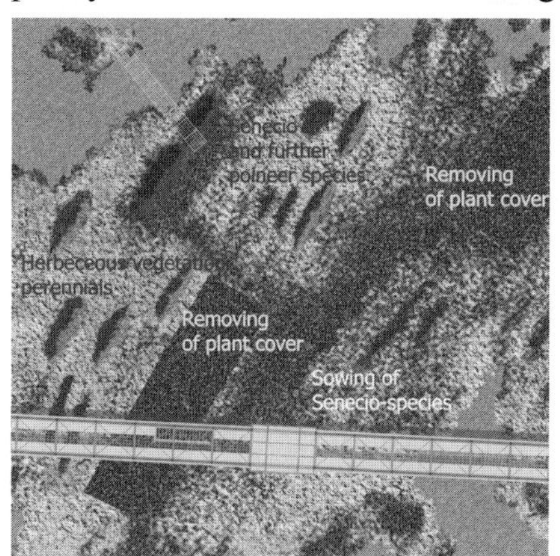

Fig. 12. Vegetation management creating a pattern of different vegetation stages

Acknowledgements

This article describes a design project elaborated at the Department of Landscape Architecture and Spatial Planning, Technical University of Munich. Thanks are due to Prof. Peter Latz.

References

Dettmar J (1999) Neue “Wildnis”. In: Dettmar J, Ganser K (eds) IndustrieNatur: Ökologie und Gartenkunst im Emscher Park. Ulmer, Stuttgart, pp 141–143

Deutsche Steinkohle AG (2000) ZT 12 SBS Abschlussberichtsplan Bergwerk Göttelborn/Reden: Bergehalde, Absinkweiher und Nachklärweiher der Tagesanlage Göttelborn. pp 11–15, 75

Grosse-Bächle L (2005) Strategies between Intervening and Leaving Room. In: Kowarik I, Körner S (eds) Urban Wild Woodlands. Springer, Berlin Heidelberg, pp 231–246

Henne SK (2005) “New Wilderness” as an Element of the Peri-Urban Landscape. In: Kowarik I, Körner S (eds) Urban Wild Woodlands. Springer, Berlin Heidelberg, pp 247–262

IndustrieKultur Saar (2000) Der Bericht der Kommission. Saarland Staatskanzlei – Stabsstelle Kultur, Saarbrücken, pp 18–22

Jochimsen M (1987) Vegetation development on mine spoil heaps—a contribution to the improvement of derelict land based on natural succession. In: Vegetation Ecology and Creation of New Environments. Proc. Intern. Symposium Tokio 1984, Tokai Univ. Press, pp 245–252

Jochimsen M (1991) Ökologische Gesichtspunkte zur Vegetationsentwicklung auf Bergehalden. In: Wiggering H, Kerth M (eds) Bergehalden des Steinkohlenbergbaus. Vieweg, Wiesbaden, pp 155–162

Körner S (2005) Nature Conservation, Forestry, Landscape Architecture and Historic Preservation: Perspectives for a Conceptual Alliance. In: Kowarik I, Körner S (eds) Urban Wild Woodlands. Springer, Berlin Heidelberg, pp 193–220

Kowarik I (1992) Das Besondere der städtischen Flora und Vegetation. Schriftenreihe des Deutschen Rates für Landespflege 61:33–47

Kowarik I (2005) Wild urban woodlands: Towards a conceptual framework. In: Kowarik I, Körner S (eds) Urban Wild Woodlands. Springer, Berlin Heidelberg, pp 1–32

Maas S (2002) Biotopstrukturkartierung Bergwerk Reden. Im Auftrag der IndustrieKultur Saar. Saarlouis, pp 7–9

Mandelbrot B (1987) Die Fraktale Geometrie der Natur (The fractal geometry of nature). Birkhäuser, Basel Boston, pp 37–69

Punz W (1989) Zur Vegetation von Hochofenschlackenhalden. Linzer biol. Beitr. 21/1:211–228

Rebele F, Dettmar J (1996) Industriebrachen: Ökologie und Management. Ulmer, Stuttgart, pp 79–100

Natur-Park Südgelände: Linking Conservation and Recreation in an Abandoned Railyard in Berlin

Ingo Kowarik, Andreas Langer

Planning Group ÖkoCon & Planland

Introduction

The particular political situation in Berlin between 1945 and 1989 had significant effects on the development of nature in the inner city. In the western part of Berlin, urban development ran in slow motion for four decades. In contrast to other parts of war-torn Europe, here large, formerly built-up areas that had been destroyed in the war remained free of renewed development; these areas were set aside as reserves to allow for future planning with Berlin as the capital city. In four decades, natural colonization processes on numerous, often heavily fragmented areas led from herbaceous and shrub stages to wild urban woodlands. The same occurred on many railyards in West Berlin because the rights for all Berlin railyards had been given by the Allies to the Reichsbahn, whose seat was in East Berlin. This organization, controlled by East Germany, reduced train service to a minimum in West Berlin, allowing natural succession to begin on many old railyards.

The special political situation of West Berlin also made possible here, earlier than in other places, the development of specific urban-industrial ecosystems which we identify today as a particular type of nature, as "nature of the fourth kind" (see Kowarik 2005); these ecosystems have long been studied systematically by Berlin's urban ecologists (see overview in Sukopp 1990). The plans of the Berlin administration provided for the integration of many of these areas into the urban open-space system because, in the walled-in western part of the city, the availability of green spaces and opportunities for experiencing nature were seen as particularly important.

After reunification in 1989, construction began on many new wilderness areas. This reversal was a symptom of a joyful change, but meant a risk

Kowarik I, Körner S (eds) Wild Urban Woodlands.

that the social and ecological functions of inner city abandoned areas would be lost. In addition to recreation functions and ecosystem services (e.g. climate regulation, hydrologic cycling), cultural-historical functions would be affected as well. The abandoned areas, with their characteristic mosaic of the remnants of former uses and natural recolonization stages, call to mind the history of the sites, especially of the historical events that first made such new natural development possible.

The Schöneberger Südgelände, which we present in this chapter, is an exception, as its condition has been secured. Originally a desolate freight railyard, then for over four decades an almost untouched new wilderness, today it is one of the first official conservation areas in Germany in which urban-industrial nature is protected and made accessible to the public. We wish to show, with the example of the "Natur-Park Südgelände," how different goals have been united and how the conceptual and design principles have opened up access to the new wilderness.

From freight railyard to "new wilderness"

The Südgelände, approximately 18 ha, lies on the southern border of the inner city of Berlin in the district of Schöneberg-Tempelhof. It is a component of a much larger freight railyard ("Rangierbahnhof bei Tempelhof") that was built between 1880–1890. Old photographs show a desolate railyard on which trains have been shunted on a multitude of parallel tracks. Tracks for the long-distance trains as well as for the inner-city express train define the area to the east and west. From the north and the south, heavily trafficked streets adjoin the site, with the result that the Südgelände has an island-like character despite its urban location.

After train service was discontinued in 1952, the Südgelände was mostly, but not entirely, abandoned. A large hall was still used for repairing the train cars, so access had to remain open. Trains were still shunted on a few tracks for a few years. On the majority of the site, however, natural development began to take place, which, by 1981, had led to a richly structured mosaic of dry grasslands, tall herbs, shrub vegetation and individual woodlands. Table 1 illustrates that between 1981 and 1991, the proportions of herbaceous vegetation and vegetation dominated by woody species had reversed. In only 10 years, the area of woodlands had doubled from 37 to 70%. Pioneer species predominate, especially the native *Betula pendula* and the North American *Robinia pseudoacacia.*

A study of the vegetation types showed that both the herbaceous and the woody vegetation are richly structured (Asmus 1981; Kowarik and Langer

1994) and provide habitats for a multitude of plant and animal species (Table 2). Rare and threatened species are found primarily in the dry grasslands and only rarely in the woody vegetation. A large proportion of the vegetation is typical of cities and differs greatly from the species composition in the rural surroundings. Among the woodlands, there are substantial differences between stands of native and non-native species. In the birch and poplar stands, a convergent development to forest communities that approach the original, widely distributed oak–pine forests is becoming apparent. In the black locust stands, on the contrary, a divergent development can be noted that can be traced back to a combination of properties of black locust that the native trees don't have at their disposal. Nitrogen fixation promotes the establishment of more demanding species (*Acer platanoides*, *A. pseudoplatanus*), and clonal growth allows black locust to regenerate within its own stands, so that it is unlikely to be entirely driven out by other trees (Kowarik 1992, 1996a, b). At least in these stands it is foreseeable that the new wilderness will be very clearly differentiated over the long term from the original communities that occurred in the Berlin area.

Table 1. Decline in herbaceous vegetation and increase in woody vegetation over a ten-year period on Berlin's Südgelände (after Kowarik and Langer 1994, data from Asmus 1981 and Kowarik et al. 1992)

	1981	1991
Area of research (ha)	22.4	20.0
Investigated vegetation cover (ha) (= 100%)	21.6	19.1
Herbaceous vegetation (%)	63.5	30.9
Woody vegetation (%)	36.5	69.1
Dominated by:		
Robinia pseudoacacia (%)	11.2	21.3
Betula pendula (%)	13.7	23.8
Betula pendula & *Populus tremula* (%)	?	5.3
Populus tremula (%)	1.3	2.3
Acer platanoides, A. pseudoplatanus (%)	0.2	1.4
Others	10.1	15.0

From new wilderness to nature park

The development of new wilderness took place at the Südgelände nearly unnoticed for a long time due to the inaccessibility of the site. Plans to completely clear the vegetation in order to erect a new freight train station led, at the beginning of the 1980s, to strong protests and to the founding of an NGO which has worked since then to preserve the Südgelände as a nature area. As a result of these efforts, a number of studies were undertaken that demonstrated the high species richness and the presence of rare species at the Südgelände (Table 2). At the end of a very changeful planning process (details in Mohrmann 2002), it was determined that the Südgelände would be set aside and developed as a nature park as a compensatory measure for new railyards in the inner city area. After a preliminary study (Kowarik et al. 1992), the Grün Berlin Park und Garten GmbH, a semi-public corporation for the development of prominent green-space projects in Berlin, commissioned the planning group ÖkoCon & Planland with the design of the nature park. After an implementation period, which was financed with funds from the government of Berlin as well as the Allianz Umweltstiftung (Allianz Foundation for Sustainability), the nature park was opened to the public in May 2000. The area has been legally set aside as the Schöneberger Südgelände landscape and nature conservation area.

Table 2. Species richness of the Schöneberger Südgelände (sources: Kowarik et al. 1992, Prasse and Ristow 1995, Saure 2001, Dahlmann pers. comm.)

	n
Vascular plants	366
Breeding birds	28
Macrofungi	49
Grasshoppers and crickets	14
Spiders	57
Wild bees and wasps	208

Challenges and approaches of the master plan

The master plan for the Natur Park Südgelände had to find planning solutions for two classic conflicts that likely arise frequently in the development of urban woodlands.

The "conservation versus recreation" conflict

The species diversity of the Südgelände (Table 2) has, in principle, developed without human intervention. The dry grasslands, in which most of the rare species are found, have emerged on nutrient-poor anthropogenic soils and are not suited to being trampled. If the small clearings of the grasslands are made accessible to visitors, eutrophication and trampling will foreseeably lead to a decline of most of the rare species. Excluding visitors, however, contradicts the goal of urban nature conservation, which is, above all, to promote natural experiences for urban residents (Auhagen and Sukopp 1983). Keeping in mind the general lack of public acceptance for nature conservation in Germany (Körner 2005, there is all the more need in urban nature conservation to combine social functions with species conservation functions.

The "wilderness versus biodiversity" conflict

In general, species diversity is greater in the earlier and middle stages of succession than in later woodland stages. This is true for the Südgelände as well with one small exception. The 40- to 50-year-old black locust stands have shown themselves to be astoundingly rich in plants, ground beetles, and spiders (Kowarik 1992; Platen and Kowarik 1995). Rare and threatened species of plants as well as hymenoptera, however, are predominantly found in the dry grasslands (Prasse and Ristow 1995; Saure 2001). A substantial increase in woodlands would emphasize the wilderness character of the Südgelände, but would also lead to a decline in the characteristic species and communities of the open landscapes.

Thus, the master plan for the Natur-Park Südgelände had to address two challenges: first, how to open the site to the public without endangering the rich local flora and fauna, and second, how to respond to the natural vegetation dynamics that would, in a short time, lead to a complete dominance of woodlands.

The model of culture and wilderness

The approach of the master plan was based on the model of simultaneity of culture and wilderness, of distance and nearness of the visitor. To implement this, a concept of zoned spaces was created in which natural and social processes were partially controlled and partially left to their own dynamics. With this approach, different goals could be combined with one another.

- In some areas, uncontrolled development of the new wilderness is allowed, without influence on the species composition. In this way, the important role of non-native species in the vegetation of the Südgelände and as a characteristic of urban vegetation was expressly accepted.

- In other areas, the open landscapes are maintained, within which succession is to be controlled through maintenance. The goal is to maintain habitats for the characteristic, and often rare species of the grasslands and other non-woody vegetation communities. In these areas, remnants of the earlier railway uses should remain at least partly recognizable. The open areas allow the underlying cultural layer of the old railyard to be easily perceived, which contrasts distinctly with the naturally derived wilderness character of the woodlands.
- In a large part of the park, the visitor may move about completely freely. A newly created path system should open the site to visitors who otherwise would have no access to the urban wilderness of an abandoned railyard.

Implementation of the model

Figure 1 shows the Natur Park Südgelände today, after significant implementation of the master plan. A few new elements were added to our planning during the implementation phase, including additional paths and the integration of works of art in the nature park.

Access concept

Starting from the park's main entrance at the S-Bahn station Priesterweg, a path system was developed that is based, fundamentally, on the linear structure of the earlier railyard. Here train tracks were made into paths (Fig. 2). Existing ramps and underpasses that once served for crossing the tracks were used to establish the path system on three different levels. Through this inclusion of the third dimension, quite different views of the area result. Because the vegetation was maintained between paths that are very near to each other, the area is perceived as larger than it actually is. A few new connections make circular routes possible.

The nature conservation area in the middle of the Südgelände is accessed by a path as well, this one, however, runs as a raised walkway 50 cm above the vegetation while following the old tracks for the most part (Fig. 3). Its design was the result of the work of the artists' group Odious. Through this new path typology it is clear to the visitor that the nature conservation area, in contrast to the rest of the Südgelände, should not be accessed off of the paths.

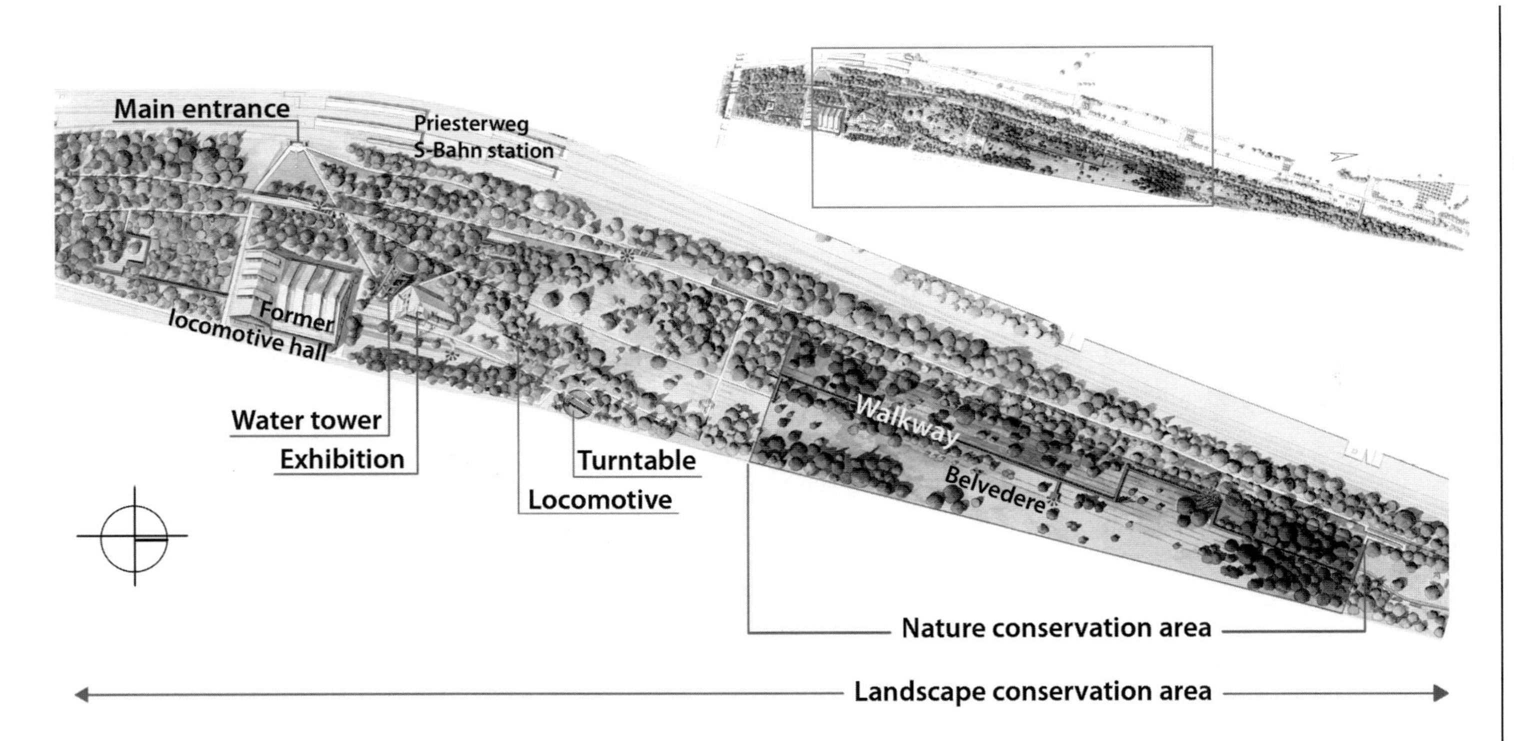

Fig. 1. Plan of the Natur-Park Südgelände (above); enlarged central section (below). During the implementation of ÖkoCon & Planland's master plan further elements were added, mainly works of art of the Odious group (Illustration: M. Ley and K. Zwingmann, Berlin, on behalf of the Senate Department of Urban Development Berlin)

Fig. 2–4. Existing train tracks are used for the path system in the Natur Park Südgelände (above). The nature conservation area in the center of the Südgelände is accessed by a raised walkway (center, below) from which the dry grasslands and the uncontrolled wilderness of the urban woodlands (below) can be perceived without leaving the path. The lower picture shows a stand of the non-native *Robinia pseudoacacia*, which has been entirely left to the natural processes of the forest dynamics.

Fig. 5–10. Creative tension between the natural dynamics of the Südgelände and the relics of the railway industry as well as the new artistic elements. The reforestation of the clearings (middle right) and the groves (below left) is prevented with maintenance measures. The picture below right, shows an art installation of the Odious group.

Definition of a room typology

In order to make clear, in accordance with the general model, the transformation from railyard to urban wilderness over time, the natural dynamics of some areas are arrested. In this way, three types of spaces or "rooms" were defined: "clearings" are to be kept free of shrubs over the long term. Stands that are light and open are to be maintained as "groves," while in the "wild woods" the natural dynamics can proceed fully unfettered.

The spatial determination of the three types was carried out according to nature conservation and landscape aesthetic criteria. In addition to the presence of rare species and communities, lines of sight within and outside of the site were considered as well as relics of the old rail industry and particularly attractive vegetation and individual trees. Before the opening of the nature park, plants were removed from overgrown areas that were once clearings or groves in order to create the predefined room structure. The open character of these rooms is ensured through long-term maintenance (mowing, removal of trees) by the Berlin nature conservation authority. The spread of the "wild woods" into the other spaces is prevented through maintenance measures. In the interior of the woodlands, natural processes proceed undisturbed (Fig. 4).

Nature conservation and recreation

Most of the Natur Park Südgelände is protected; the core area has been designated as a nature reserve (3.2 ha) and the rest as a landscape conservation area (12.9 ha). In the nature reserve, species conservation takes priority. The targeted species are, above all, insects of open habitats and plants of the dry grasslands, such as several rare hawkweed (*Hieracium*) species. The clearings, which may not be entered by the public, are cared for in such a way as to give the characteristic species of the open landscape a chance to survive. The landscape conservation area is to be fully accessible. Dangerous areas that aren't visible (e.g. shafts) were secured before the park was opened. The defined rooms are stabilized through maintenance measures. In the landscape conservation area, an attractive landscape image is more the goal than species conservation.

Since its opening in May 2000, the Natur Park Südgelände has proved to be very attractive to visitors. Estimates start at 50,000 visitors per year. A long-term exhibition on the history and nature of the Südgelände in one of the old train buildings had approximately 6,000 visitors in 2003.

Two cultural layers

To maintain the visibility of the remnants of the railway history in the face of the powerful natural dynamics, selected railway relics such as the signals and the old turntable were restored. The many paths set in the old tracks are a permanent reminder of the cultural foundation of the nature development of the Südgelände. A new cultural layer has been established through the art works of the Odious group, which present a creative tension with the developing wilderness as well as with the relics of the railway (Fig. 10). The water tower was secured as a landmark of the Südgelände (Fig. 5), old buildings were surrendered to a controlled decay or are used for the exhibition or as studios for the artists.

Conclusion

Is the Natur-Park Südgelände a good example of a successful integration of urban wilderness into the open-space system of a metropolis? What speaks in its favor is the simple fact that this kind of nature development has indeed been successfully safeguarded despite substantial competition for use in the reunited German capital. The contrast between dynamic nature and the remnants of the railway industry heritage is fascinating to all visitors. Unfettered wilderness development is always taking place in parts of the Südgelände. Through the spatially differentiated maintenance plan, the earlier and middle stages of nature development are maintained and thereby the diversified vegetation complexes are maintained in the long term. The species targeted for nature conservation profit as well from the maintenance measures. The public acceptance of the nature park is extremely high.

The original railway wilderness has, however, clearly been affected by design interventions in the form of the new path system, the maintenance and the art objects. Has this destroyed the original uniqueness, the "wilderness" of the Südgelände? Certainly the character of the site has changed. The few who earlier had discovered the Südgelände on their own recognize the contrast very clearly. To wake Sleeping Beauty, however, also means to open the urban wilderness to a multitude of visitors who did not have an inherent sympathy for the nature of urban abandoned areas. That such access, even designed access, satisfies a need for wilderness has been shown in studies such as the one by Bauer (2005).

The wild urban woodlands of the Industriewald Ruhrgebiet (the Industrial Forest of the Ruhr) have been made accessible very successfully through landscape architectural means and through works of art (Dettmar 2005). The Südgelände, however, is much smaller than most of the abandoned areas of the Ruhr, so the proportion of designed elements is greater here and perhaps sometimes competes with the natural processes that are characteristic of the area. Arrangements should therefore continue to be fine-tuned (Kowarik et al. 2004). Taken together, however, there is a great deal of evidence that the Natur Park Südgelände has been successful in bringing humans living in urban neighborhoods a step closer to biodiversity in its characteristic urban expression.

Acknowledgements

Our thanks are due to the Senate Department of Urban Development Berlin for providing Fig. 1 and to Kelaine Vargas for the translation.

References

Asmus U (1981) Vegetationskundliches Gutachten über das Südgelände des Schöneberger Güterbahnhofs. Unpublished research report on behalf of the Senat für Bauen und Wohnen, Berlin

Auhagen A, Sukopp H (1983) Ziel, Begründungen und Methoden des Naturschutzes im Rahmen der Stadtentwicklungspolitik von Berlin. Natur und Landschaft 58(1):9-15

Bauer N (2005) Attitudes towards Wilderness and Public Demands on Wilderness Areas. In: Kowarik I, Körner S (eds) Urban Wild Woodlands. Springer, Berlin Heidelberg, pp 47–66

Dettmar J (2005) Forests for Shrinking Cities? The Project "Industrial Forests of the Ruhr". In: Kowarik I, Körner S (eds) Urban Wild Woodlands. Springer, Berlin Heidelberg, pp 263–276

Körner S (2005) Nature Conservation, Forestry, Landscape Architecture and Historic Preservation: Perspectives for a Conceptual Alliance. In: Kowarik I, Körner S (eds) Urban Wild Woodlands. Springer, Berlin Heidelberg, pp 193–220

Kowarik I (1992) Einführung und Ausbreitung nichteinheimischer Gehölzarten in Berlin und Brandenburg. Verh Bot Ver Berlin Brandenburg, Beiheft 3:1–188

Kowarik I (1996a) Primäre, sekundäre und tertiäre Wälder und Forsten. Mit einem Exkurs zu ruderalen Wäldern in Berlin. Landschaftsentwicklung und Umweltforschung 104:1-22

Kowarik I (1996b) Funktionen klonalen Wachstums von Bäumen bei der Brachflächen-Sukzession unter besonderer Beachtung von Robinia pseudoacacia. Verh Ges Ökol 26:173–181

Kowarik I (2005) Wild urban woodlands: Towards a conceptual framework. In: Kowarik I, Körner S (eds) Urban Wild Woodlands. Springer, Berlin Heidelberg, pp 1–32

Kowarik I, Langer A (1994) Vegetation einer Berliner Eisenbahnfläche (Schöneberger Südgelände) im vierten Jahrzehnt der Sukzession. Verh Bot Ver Berlin Brandenburg 127:5-43

Kowarik I, Aey W, Bruhn K, Fietz M, Koehler N, Langer A, Meissner J, Machatzi B, Prasse R, Ristow M, Röttger N, Saure C, Scholler M, Siederer W (1992)

Naturpark Südgelände: Bestand, Bewertung, Planung. Unpublished report by order of Bundesgartenschau Berlin 1995 GmbH, Berlin

Kowarik I, Körner S, Poggendorf L (2004) Südgelände: Vom Natur- zum Erlebnis-Park. Garten und Landschaft 114(2):24–27

Mohrmann R (2002) Beispiel Naturpark Schöneberger Südgelände in Berlin. In: Auhagen A, Ermer K, Mohrmann R (eds) Landschaftsplanung in der Praxis, Ulmer, Stuttgart, pp 328-354

Platen R, Kowarik I (1995) Dynamik von Pflanzen-, Spinnen- und Laufkäfergemeinschaften bei der Sukzession von Trockenrasen zu Gehölzgesellschaften auf innerstädtischen Brachflächen in Berlin. Verh Ges f Ökol 24:431–439

Prasse R, Ristow M (1995) Die Gefäßpflanzenflora einer Berliner Güterbahnhofsfläche (Schöneberger Südgelände) im vierten Jahrzehnt der Sukzession. Verhandlungen des Botanischen Vereins von Berlin und Brandenburg 128:165-192

Saure C (2001) Das Schöneberger Südgelände. Ein herausragender Ruderalstandort und seine Bedeutung für die Bienenfauna (Hymenoptera, Apoidea). Berliner Naturschutzblätter 35: 17-29

Sukopp H (ed) Stadtökologie. Das Beispiel Berlin. Reimer, Berlin